SpringerBriefs in Electrical and Computer Engineering

SpringerBriefs in Speech Technology

Series Editor

Amy Neustein

For further volumes:
http://www.springer.com/series/10043

Editor's Note

The authors of this series have been hand selected. They comprise some of the most outstanding scientists—drawn from academia and private industry—whose research is marked by its novelty, applicability, and practicality in providing broad-based speech solutions. The Springer Briefs in Speech Technology series provides the latest findings in speech technology gleaned from comprehensive literature reviews and *empirical investigations* that are performed in both laboratory and *real life* settings. Some of the topics covered in this series include the presentation of real life commercial deployment of spoken dialog systems, contemporary methods of speech parameterization, developments in information security for automated speech, forensic speaker recognition, use of sophisticated speech analytics in call centers, and an exploration of new methods of soft computing for improving human-computer interaction. Those in academia, the private sector, the self service industry, law enforcement, and government intelligence are among the principal audience for this series, which is designed to serve as an important and essential reference guide for speech developers, system designers, speech engineers, linguists, and others. In particular, a major audience of readers will consist of researchers and technical experts in the automated call center industry where speech processing is a key component to the functioning of customer care contact centers.

Amy Neustein, Ph.D., serves as editor in chief of the *International Journal of Speech Technology* (Springer). She edited the recently published book *Advances in Speech Recognition: Mobile Environments, Call Centers and Clinics* (Springer 2010), and serves as quest columnist on speech processing for Womensenews. Dr. Neustein is the founder and CEO of Linguistic Technology Systems, a NJ-based think tank for intelligent design of advanced natural language-based emotion detection software to improve human response in monitoring recorded conversations of terror suspects and helpline calls.

Dr. Neustein's work appears in the peer review literature and in industry and mass media publications. Her academic books, which cover a range of political, social, and legal topics, have been cited in the Chronicles of Higher Education and have won her a pro Humanitate Literary Award. She serves on the visiting faculty of the National Judicial College and as a plenary speaker at conferences in artificial intelligence and computing. Dr. Neustein is a member of MIR (machine intelligence research) Labs, which does advanced work in computer technology to assist underdeveloped countries in improving their ability to cope with famine, disease/illness, and political and social affliction. She is a founding member of the New York City Speech Processing Consortium, a newly formed group of NY-based companies, publishing houses, and researchers dedicated to advancing speech technology research and development.

K. Sreenivasa Rao • Anil Kumar Vuppala

Speech Processing in Mobile Environments

 Springer

K. Sreenivasa Rao
Indian Institute of Technology
Kharagpur, West Bengal, India

Anil Kumar Vuppala
International Institute of Information
Technology
Hyderabad, Gachibowli, India

ISSN 2191-8112   ISSN 2191-8120 (electronic)
ISBN 978-3-319-03115-6   ISBN 978-3-319-03116-3 (eBook)
DOI 10.1007/978-3-319-03116-3
Springer Cham Heidelberg New York Dordrecht London

Library of Congress Control Number: 2013956607

Printed on acid-free paper

Springer is part of Springer Science+Business Media (www.springer.com)

Preface

Robust speech systems in mobile environment have gained a special interest in recent years in order to enable access to remote voice-activated services. In this context, three major challenges that need to be considered are: varying background conditions, speech coding, and transmission channel errors. In this book, we focus on improving the recognition performance of speech systems in the presence of speech coding and background noise conditions by using vowel onset points (VOPs) as anchor points. VOP is an important event in speech production, and it is defined as the instant at which the onset of vowel takes place. Speech coders considered in this work are GSM full rate (ETSI 06.10), GSM enhanced full rate (ETSI 06.60), CELP (FS-1016), and MELP (TI 2.4 kbps).

The major works presented in this book are:

- Methods are proposed for the detection of VOPs in the presence of speech coding and background noise conditions.
- A two-stage hybrid approach based on hidden Markov models (HMMs) and support vector machines (SVMs) is proposed for improving the performance of consonant-vowel (CV) recognition system.
- Two-stage VOP detection method is proposed for spotting CV units from continuous speech.
- Combined temporal and spectral preprocessing methods are explored to improve the performance of CV recognition system under background noise.
- A method based on VOPs is proposed to improve the performance of speaker identification (SI) system in the presence of coding.
- A method is proposed for nonuniform time scale modification using VOPs and instants of significant excitation.

Some important conclusions drawn out of this work are: (i) Performance of the proposed VOP detection method based on spectral energy in the glottal closure region is found to be better compared to existing methods under clean, coded and noisy conditions. (ii) Performance of the proposed two-stage hybrid CV recognition approach has shown significant improvement compared to other approaches, under clean, coded, and noisy conditions. (iii) Performance of CV recognition system under background noise is improved by using combined temporal and spectral processing-based preprocessing method. (iv) Proposed two-stage VOP detection method used for spotting CV segments from continuous speech has found to be efficient in minimizing the missing and spurious VOPs. (v) In the presence of coding, performance of SI system is improved by using features extracted from steady vowel speech segments. Improvement in SI system performance is mainly due to the presence of crucial speaker-specific information in the steady vowel segments of speech, even after coding. (vi) Performance of the proposed time scale modification method is superior compared to existing methods. The superior performance of the proposed method is due to the nonuniform modification of different speech segments and accurate detection of various speech segments with the help of instants of significant excitation and VOPs.

This book is mainly intended for researchers working on building robust speech systems in mobile environment. This book is also useful for the young researchers, who want to pursue the research in speech processing. Hence, this may be recommended as the text or reference book for the postgraduate level advanced speech processing course.

Many people have helped us during the course of preparation of this book. We would especially like to thank all professors of G.S. Sanyal School of Telecommunication and School of Information Technology, IIT Kharagpur for their moral encouragement and technical discussions during the course of editing and organization of the book. Special thanks to our colleagues at IIT Kharagpur for their cooperation and coordination to carry out the work. Finally, we thank all our friends and well-wishers.

West Bengal, India   K. Sreenivasa Rao
Hyderabad, India   Anil Kumar Vuppala

Contents

Acronyms

AANN	Auto-associative neural network
ACELP	Algebraic code excited linear prediction
AIR	All India Radio
CV	consonant-vowel
CELP	Code excited linear prediction
COMB-ESM	Combining excitation source, spectral peaks, and modulation spectrum
CSM	Constraint satisfaction model
CSNN	Constraint satisfaction neural network
DCT	Discrete cosine transform
DFT	Discrete Fourier transform
ES	Excitation source
EFR	Enhanced full rate
FFNN	Feed-forward neural network
FOD	First order difference
FOGD	First order gaussian difference
FR	Full rate
GC	Glottal closure
GCI	Glottal closure instants
GSM	Global system for mobile
GMM	Gaussian mixture models
HE	Hilbert envelope
HMM	Hidden Markov model
HTK	HMM tool kit
Hz.	Hertz
IDFT	Inverse discrete Fourier transform
ISE	Instant of significant excitation
kHz.	Kilo hertz
LP	Linear prediction
MFCC	Mel frequency cepstral coefficients
MELP	Mixed excitation linear prediction

MIPS	Million instructions per second
MLFFNN	Multilayer feed-forward neural network
MMSE	Minimum mean square error
MOD	Modulation spectrum
MOS	Mean opinion score
NN	Neural networks
PROP-COD	Proposed VOP detection method for coded speech
PSOLA	Pitch synchronous overlap and add
PST	Perceptually significant transient
SCV	Stop consonant-vowel
SI	Speaker identification
SMV	Selective mode vocoder
SNR	Signal-to-noise ratio
SS	Spectral subtraction
STSA	Short time spectral amplitude
SVM	Support vector machines
TD	Time domain
TSM	Time scale modification
TSP	Temporal spectral processing
UMTS	Universal mobile telecommunications system
VoIP	Voice over internet protocol
VOP	Vowel onset point
VOT	Voice onset time
VT	Vocal track
WSOLA	Waveform similarity overlap-and-add
ZFF	Zero frequency filter

Chapter 1
Introduction

Abstract The rapid growth of mobile users is creating great deal of interest in the development of robust speech systems in mobile environment. Some of the new and exciting services enabled by speech systems in mobile environment are: speech interface to the mobile devices, information retrieval through mobile devices, voice-based person authentication, and forensic investigation. Issues involved in adapting the present speech processing technology to mobile systems are: effect of varying background noise, degradations introduced by the speech coders, and errors introduced due to transmission impairments. In this work, the major focus is on improving the recognition performance of speech systems in the presence of speech coding and background noise by using vowel onset points (VOPs). This chapter provides the overall objective of the present work and scope of the book. The chapter-wise organization and evolution of ideas related to present work are given at the end of this chapter.

1.1 Introduction

Speech is produced as a sequence of changes and those are known as events. For example, in speech there are phonetic events and acoustic events. Any change which can be attributed to the activity of the speech organs is a phonetic event. For example, voicing and closure are phonetic events [1–3]. Any feature which is present in the acoustic signal is an acoustic event. For example, burst, friction, and voice onset time (VOT) are acoustic events. From the perception point of view, events and regions around them are known to contain important information [4]. Conventional block processing approach uses fixed frame size (20–30 ms) to extract information, and it does not use knowledge of events. Vowel onset point (VOP) is one of the important event in speech production. The VOP is defined as the instant at which the onset of vowel takes place in the speech signal. The significance of VOP can be observed in speech applications like (1) recognition of Consonant-Vowel (CV) units, (2) spotting CV segments in continuous speech, (3) speaker recognition,

K.S. Rao and A.K. Vuppala, *Speech Processing in Mobile Environments*, SpringerBriefs in Electrical and Computer Engineering, DOI 10.1007/978-3-319-03116-3_1,
© Springer International Publishing Switzerland 2014

(4) speech rate manipulation, and (v) enhancement of speech [4–6]. Accuracy in the detection of VOP is vital for these applications. Therefore in this work, we propose accurate VOP detection methods under clean and degraded environments.

1.2 Objective of the Book

At the signal level, robust information is present in speech around glottal closure and VOP events [4]. The objective of this work is to illustrate the significance of accurate VOP detection for speech processing in mobile environment. Existing VOP detection methods are suffering with poor detection accuracy. Speech signal during glottal closure regions exhibits high signal-to-noise ratio (SNR) characteristics. Hence, processing glottal closure regions may be useful for accurate detection of VOPs under clean and degraded conditions. The knowledge of VOP events is used in the following studies:

- Recognition of CV units in the presence of speech coding and background noise.
- Spotting and recognition of CV units from continuous speech.
- Speaker identification (SI) in the presence of coding.
- Nonuniform time scale modification (TSM)

1.3 Organization of the Book

The evaluation of ideas presented in this book are listed in Table 1.1. The rest of the book is organized as follows:

Chapter 2: Background and Literature Survey—discusses the state-of-the-art methods for VOP detection and speech systems in mobile environment. Existing approaches for CV recognition in Indian languages and time scale modification are also discussed in this chapter.

Chapter 3: Vowel Onset Point Detection from Coded and Noisy Speech—presents the proposed VOP detection methods for coded and noisy speech. Performance of the proposed VOP detection methods is compared with existing method which uses the combination of evidences from excitation source, spectral peaks, and modulation spectrum.

Chapter 4: Consonant-Vowel Recognition in the Presence of Coding and Background Noise—presents the recognition performance of the CV units in the presence of coding and background noise by using proposed two-stage hybrid approach. Proposed CV recognition approach uses the combination of complimentary evidences from support vector machine (SVM) and hidden Markov model (HMM) to improve recognition performance. Impact of accuracy in the proposed VOP detection method is studied on recognition performance of CV units by using proposed CV recognition approach in the presence of coding. Further,

Table 1.1 Evolution of ideas presented in the book

- Speech systems in mobile environment has become popular in recent years
- The major issues in mobile environment are background noise, speech coding, and channel errors
- Information present around VOP can be useful for speech processing in mobile environment
- Existing VOP detection methods are suffering with poor accuracy. Hence, there is a need to develop accurate VOP detection methods for both clean and degraded conditions
- Glottal closure regions are known to be high SNR regions in speech. Therefore, spectral energy of the speech signal present in the glottal closure region can be explored for the detection of VOPs in the presence of coding and background noise
- Crucial information of CV unit is present around VOP, and hence VOP can be used as an anchor point for deriving the relevant information for CV recognition
- As the number of CV classes are more, multistage acoustic models may perform better compared to single stage acoustic models. Hence, we explored a two-stage hybrid approach for improving the recognition performance of CV units
- HMMs and SVMs are trained using different modalities, hence they can provide complementary evidence. Therefore, at each stage of the proposed CV recognition method, complementary evidences from SVM and HMM are combined for enhancing the CV units recognition performance
- Combined TSP methods are known to be useful for improving the performance in speech enhancement and speaker recognition tasks. In this work, we explored its usefulness in speech recognition task
- VOPs can be used for spotting CV units from continuous speech. Hence, two-stage accurate VOP detection method is proposed for spotting and recognition of CV units from continuous speech
- Since speaker-specific characteristics are preserved in steady vowel segments of speech even after coding, the features extracted from these steady vowel regions can be used to improve the SI performance in the presence of coding. Hence, a method is proposed to determine the steady vowel region from the speech signal by using VOPs and epochs
- Due to unique articulatory and production constraints associated with each type of vowel during slow and fast speech, vowel segments are expanded and compressed nonuniformly based on the type of vowel. Therefore, a nonuniform TSM method is proposed by using VOPs and epochs

combined temporal spectral processing (TSP)-based preprocessing methods are used to improve the recognition performance of CV units under background noise.

Chapter 5: Spotting and Recognition of Consonant-Vowel Units from Continuous Speech—discusses about the need for accuracy in the detection of VOP for CV recognition in continuous speech. Proposed two-stage VOP detection method to reduce the spurious VOPs and improve the accuracy of genuine VOPs is presented in this chapter.

Chapter 6: Speaker Identification and Time Scale Modification Using VOPs—focuses on the application of proposed VOP detection methods for improving the SI performance in the presence of coding, and nonuniform time scale modification (TSM). Proposed speaker identification system is developed using features extracted from steady vowel region. Steady vowel regions are determined by using vowel onset points and epochs. Further, proposed nonuniform TSM method is presented for slow down and speed up the speech.

Chapter 7: Conclusions—summarize the contributions of the book and discuss the scope for future investigation.

Chapter 2
Background and Literature Review

Abstract This chapter provides the systematic review of the existing approaches for vowel onset point detection, speech systems in mobile environment, Consonant-Vowel (CV) recognition in Indian languages, and time scale modification (TSM). In addition to providing the review of above-mentioned topics, authors have discussed about the short comings present in the existing approaches and derived the motivation and scope of the present work.

This chapter discusses about the state of the art related to the contents of this book. Authors have provided detailed explanation for the existing VOP detection methods, CV recognition systems, and time scale modification methods, which are later used for comparing the performance with the proposed methods and systems. The chapter is organized as follows: Sect. 2.1 reviews existing methods for VOP detection. Section 2.2 briefly reviews the state-of-the-art speech and speaker recognition systems in mobile environment. Section 2.3 presents the review of CV recognition in Indian languages. Section 2.4 reviews the existing approaches for TSM. Section 2.5 summarizes the review and the major issues addressed in this book.

2.1 Approaches for Detection of Vowel Onset Points

There are various methods available in literature for the detection of VOPs [7–17]. The method presented in [7] detects VOPs based on rapid increase in the vowel strength. The vowel strength is calculated using the difference in the energy of the peaks and their corresponding valleys in the amplitude spectrum. This method requires detection of unvoiced and voiced regions in a speech signal. The method for VOP detection presented in [8] uses a product function generated by using wavelets. The values of product function for vowel regions are much larger than consonant regions. The methods presented in [9–11] use a hierarchical neural

K.S. Rao and A.K. Vuppala, *Speech Processing in Mobile Environments*, SpringerBriefs in Electrical and Computer Engineering, DOI 10.1007/978-3-319-03116-3_2,
© Springer International Publishing Switzerland 2014

network, multilayer feed-forward neural network (MLFFNN), and auto-associative neural network (AANN) models, respectively, for the detection of VOPs. These models are trained by using the trends in the speech signal parameters at the VOPs. The VOP detection using Hilbert envelope of excitation source information is presented in [12]. The acoustic cues such as formant transition, epoch intervals, strength of instants, symmetric Itkura distance, and ratio of signal energy to residual energy are explored in [4, 5] for the detection of VOP events in different categories of CV units.

The voice onset time (VOT) is the time delay between the burst onset and the start of periodicity, when it is followed by a voiced sound. In [13], automatic VOT is detected using phone model-based methods with forced alignment. VOT detection using reassignment spectra is presented in [14]. In [18], voice onset time detection method is presented for unvoiced stops (/p/, /t/ and /k/) using the nonlinear energy tracking algorithm (Teager energy operator). In [19], Bessel features are used for determining the voice onset time for stop consonant vowel units such as /ka/, /Ta/, /ta/, and /pa/.

Combination of the evidence from excitation source, spectral peaks, and modulation spectrum (COMB-ESM) has been explored in [15] for the detection of VOPs. Each of these evidence carries complementary information with respect to VOPs. The performance of COMB-ESM method is superior compared to existing methods. Hence, in this book COMB-ESM method is used for comparing the performance of the proposed VOP detection methods. Following subsection describes the details of the COMB-ESM method for VOP detection [15].

2.1.1 VOP Detection Using Excitation Source, Spectral Peaks, Modulation Spectrum, and Their Combination

2.1.1.1 VOP Detection Using Excitation Source Information

VOP detection using excitation source information is carried out in the following sequence of steps. Determine the Hilbert envelope (HE) of linear prediction (LP) residual (also known as excitation source) of speech signal. Smooth the HE of the LP residual by convolving with a Hamming window of size 50 ms. The change at the VOP present in the smoothed HE of the LP residual is further enhanced by computing its slope using first-order difference (FOD). These enhanced values are convolved with the first order Gaussian difference (FOGD) operator, and the convolved output is the VOP evidence using excitation source. VOP evidence using excitation source for speech signal /"Don't ask me to carry an"/ is shown in Fig. 2.1b.

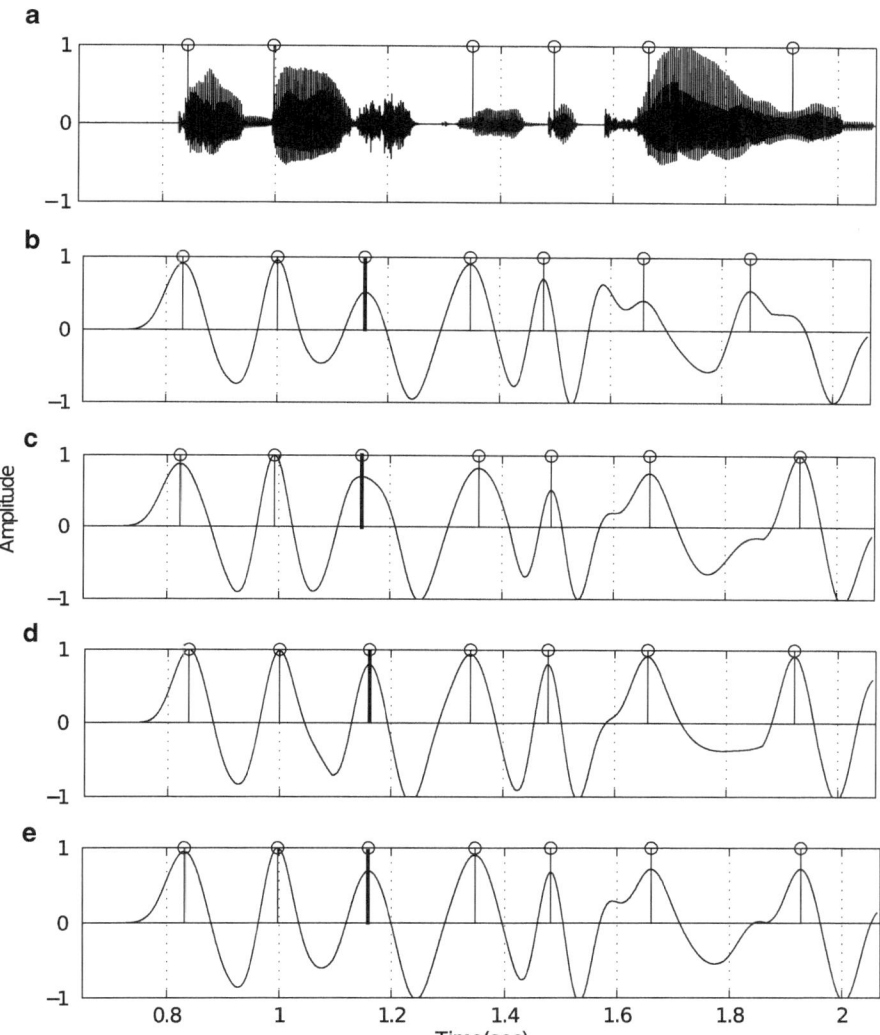

Fig. 2.1 VOP detection using combination of three evidences for a speech utterance /*"Don't ask me to carry an"*/. (**a**) speech signal, VOP evidence plots for (**b**) excitation source, (**c**) spectral peaks, (**d**) modulation spectrum, and (**e**) COMB-ESM method

2.1.1.2   VOP Detection Using Spectral Peaks Energy

VOP detection using the spectral peaks energy is carried out in the following sequence of steps. The speech signal is processed in blocks of 20 ms with a shift of 10 ms. For each block, a 256-point discrete Fourier transform (DFT) is computed, and the ten largest peaks are selected from the first 128 points. The sum of these spectral peaks is plotted as a function of time. The change at the VOP available in the spectral peaks energy is further enhanced by computing its slope using FOD. These enhanced values are convolved with FOGD operator. The convolved output is the VOP evidence using spectral peaks energy. VOP evidence plot using spectral peaks energy for speech signal /"*Don't ask me to carry an*"/ is shown in Fig. 2.1c.

2.1.1.3   VOP Detection Using Modulation Spectrum Energy

Slowly varying temporal envelope of speech signal can be represented by using modulation spectrum. VOP detection using modulation spectrum energy is carried out in the following sequence of steps. The temporal envelope of speech is dominated by low-frequency components. The VOP evidence due to modulation spectrum is derived by passing the speech signal through a set of critical band pass filters and summing the components corresponding to 4–16 Hz. The change at the VOP available in the modulation spectrum energy is further enhanced by computing its slope using FOD. These enhanced values are convolved with FOGD operator and the convolved output is the VOP evidence using modulation spectrum energy. VOP evidence using modulation spectrum energy for speech signal /"*Don't ask me to carry an*"/ is shown in Fig. 2.1d.

2.1.1.4   VOP Detection Using COMB-ESM Method

Each of the above three methods uses complementary information about the VOP, and hence they are combined for the enhancement of VOP detection performance. In combined method, the evidences from excitation source, spectral peaks, and modulation spectrum energies are added sample by sample. VOP detection using individual and combination of all three evidences for speech signal /"*Don't ask me to carry an*"/ is shown in Fig. 2.1.

Figure 2.1a shows the speech signal with manually marked VOPs for an utterance /"*Don't ask me to carry an*"/. Figure 2.1b–d shows the VOP evidence corresponding to excitation source, spectral peaks, and modulation spectrum, respectively. Figure 2.1e shows the VOP evidence by combining the evidence. The peaks in the combined VOP evidence signal (Fig. 2.1e) are marked as the VOPs obtained from COMB-ESM method. From Fig. 2.1, it is observed that a spurious VOP is present in third position in all VOP evidence plots. The performance of COMB-ESM method for VOP detection is around 96 % within 40 ms deviation and only 45 % within 10 ms deviation [15]. A summary of the discussion related to the detection of VOP is given in Table 2.1.

Table 2.1 Summary of the review of VOP detection methods

- Existing methods for VOP detection have low accuracy
- Most of the existing VOP detection methods are based on block processing of speech signals
- Information present in glottal closure regions may also be used for the detection of the VOP events in the presence of degradations such as coding and background noise

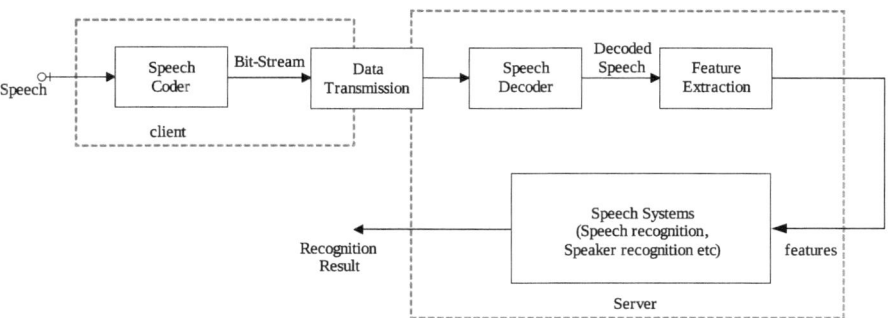

Fig. 2.2 Block diagram for network-based speech system

2.2   Speech Processing in Mobile Environment

In mobile environment, speech systems are developed in three different configurations. They are: (1) embedded speech systems (client-based), (2) network speech systems (server-based), and (3) distributed speech systems (client-server-based) [20–24]. These configurations are characterized according to the location where processing is taking place. In embedded speech systems, the speech task is performed in the terminal device itself. Due to cost-sensitive nature of the terminal device, constraints are imposed on computational and memory resources. Therefore, this approach is aimed to limited vocabulary applications. In network speech systems, speech is transmitted to remote server over a communication channel, and speech task is performed at the server. In distributed speech systems, features required for speech task are extracted from the speech at the client side, and task is performed at the remote server. The applications such as large continuous speech recognition in mobile environment is possible through network and distributed based speech systems. In this book, network-based configuration is considered for analyzing the proposed methods for speech and speaker recognition. The block diagram for network-based speech system is shown in Fig. 2.2. Survey of speech recognition techniques for mobile devices is presented in [20–23, 25].

There are three major challenges in speech processing in mobile environment [21–23, 25]: (1) Degradations due to different speech coders used for speech transmission. Speech coding is a compact way of representing speech by exploiting speech production and perception characteristics. In the process of speech coding, speech and speaker-specific information present in speech will be degraded.

(2) Effect of varying background conditions in mobile environment on the performance of speech systems. Background conditions like crowd of people, vehicle, restaurant, street, etc. are common in mobile environment, and they will degrade the performance of speech systems. (3) Effect of wireless channels on the performance of speech systems. Due to unreliable nature of radio-frequency channel, transmission errors will affect the performance of speech systems. Distortions due to speech coding and channel errors like packet loss are also common issues in voice over internet protocol (VoIP). In addition to these distortions, jitter is an issue in VoIP technology.

Currently speech coders are coming in two different versions. The basic version, also called narrowband, which is mainly intended for use by GSM, and wideband (for example AMR-WB), which is mainly intended for use by Universal Mobile Telecommunications System (UMTS). UMTS is one of the third generation (3G) mobile telecommunications systems. Wideband coders uses a speech bandwidth of 50–7,000 Hz, whereas the bandwidth of narrowband AMR is 300–3,400 Hz. This gives wideband AMR a more natural speech quality. We consider different narrowband speech coders to observe coding effect on the performance of speech systems. In this work, issues related to speech and speaker recognition under coding and speech recognition under background noise are addressed. Following subsections discuss the background work related to those issues.

2.2.1 Speech and Speaker Recognition Under Coding

Speech recognition is the process of converting spoken words to a machine readable input (text). The effect of speech coders such as GSM and CELP coders on digit recognition performance by using HMM models has been discussed in [26, 27]. Juan Huerta has presented weighted acoustic models to reduce the effect of GSM full rate coder on the speech recognition performance [22]. From his study it is evident that all phonemes in a GSM-coded speech corpus are not distorted to the same extent due to coding. Alternative front-end for speech recognition in GSM networks is presented in [28]. In this approach features are extracted directly from the encoded speech to avoid source coding distortion.

Speaker recognition is the process of automatically recognizing the identity of speaker from speech. Speaker recognition can be divided into speaker identification and speaker verification. In speaker identification, the task is to identify the speaker from the speech signal. The task of a speaker verification system is to authenticate the claim of a speaker based on the test speech. In literature the effect of coding on speaker recognition performance was analyzed in two ways. In the first case features required for speaker recognition are extracted from resynthesized speech [29–31], and in the second case features are extracted directly from the codec parameters [29]. The effect of GSM (12.2 kbps), G.729 (8 kbps), and G.723.1 (5.3 kbps) coders on speaker recognition is studied in [29]. This study indicated

Table 2.2 Summary of the review of speech systems in mobile environment

- There is no systematic study carried out on speech recognition for Indian languages in mobile environment
- Combined temporal and spectral preprocessing techniques can be used for speech recognition under background noise
- Information present around VOPs may be used for improving the performance of speech systems in mobile environment

that the performance of recognition system decreases with decreasing coding rate. The effect of speech coding on automatic speaker recognition is presented in [30] with matched and mismatched training and testing conditions. Matched condition (training and testing with the same coder) shows increase in the recognition performance. In [31], performance of speaker recognition under coding is improved using score normalization. The effect of GSM-EFR coder on the performance of speaker identification is presented in [32]. In [33], SVM-based text-independent speaker identification using a linear GMM supervector kernel was presented for coded speech.

2.2.2   Speech Recognition Under Background Noise

In practical applications of automatic speech recognition, speech is often distorted by a background noise. Because of this distortion, speech features are distorted, and therefore there is a mismatch between the training (clean) and testing (noisy) conditions. This mismatch severely degrades the performance of speech recognizers [34, 35]. Various methods have been presented in the literature to overcome the effect of noise on speech recognition. These methods can be grouped under three categories based on (1) compensation of noise, (2) robust feature extraction, and (3) adaptation of models. Methods based on compensation of noise aim to enhance the noisy speech signals before feature extraction [34–39]. Such methods include spectral subtraction, minimum mean square error (MMSE), and subspace-based speech enhancement techniques [36–39]. Methods based on robustness at the feature level are designed in such a way that the proposed features are less sensitive to the noisy degraded conditions [40–46], e.g., RASTA filter [40], feature normalization [41], MMSE-based mel-frequency cepstra [42], and histogram equalization [44–46], etc. In case of model adaptation approach, the parameters of the model are modified according to the characteristics of the background noise [47–53]. Some of the popular model adaptation methods include code book mapping [47], parallel model compensation [48], noise adaptive training [51, 52], etc. A summary of the discussion related to the speech systems in mobile environment is given in Table 2.2.

2.3   Recognition of CV Units of Speech in Indian Languages

Phones, diphones, and triphones are widely used subword units for speech recognition. But recent studies reveal that syllables are the suitable subword units for speech recognition [54,55] in Indian languages. In general, the syllable-like units are of type $C^m V C^n$, where C refers to consonant, V refers to a vowel, and m and n refer to the number of consonants preceding and following the vowel in a syllable. Among these units, the CV units are the most frequently (around 90 %) occurring units [9] in Indian languages. Different regions of significant events in the production of the CV unit $/ka/$ are shown in Fig. 2.3. The major issues involved in the recognition of CV units are the large number of classes and high similarity among those classes [54–57].

Hidden Markov models (HMMs) are the commonly used classification models in speech recognition, but in [54, 55, 57] authors have reported that MLFFNNs and support vector machines (SVMs) work better for recognition of CV units in Indian languages compared to HMM. In [55], modular neural networks are used for recognition of stop consonant-vowel (SCV) units. Separate neural networks (subnets) are trained for subgroups of classes. It has been reported in [55] that the performance of the conventional modular networks is poor, and a constraint satisfaction model (CSM) is presented to improve the recognition performance of SCV units. In CSM the outputs of the subnets are combined using the constraints that represent the similarities among the SCV classes. The constraints are derived from the acoustic phonetic knowledge of the classes and the performance of the subnets. In [54], constraint satisfaction neural network models are extended for recognition of isolated CV units that correspond to all categories of consonants. Features extracted around VOPs are used for recognition of CV units. In their study, VOPs are detected using AANNs and dynamic time warping (DTW)-based methods [54]. Further, CV units are recognized from continuous speech by using SVMs.

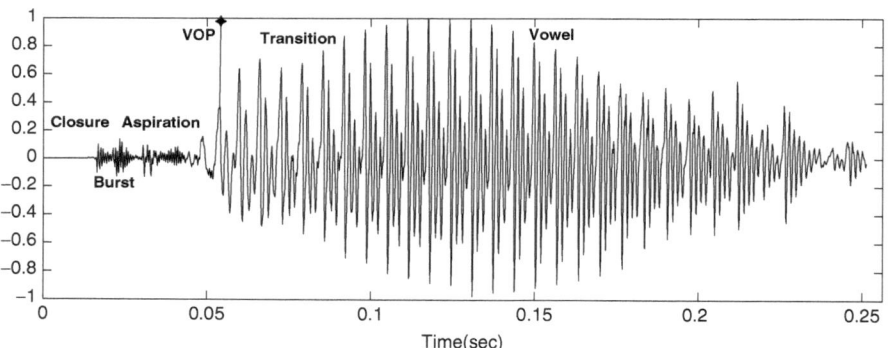

Fig. 2.3 Regions of significant events in the production of the CV unit $/ka/$

Table 2.3 Summary of the review of CV recognition in Indian languages

- VOP detection methods used in existing CV recognition systems are suffering with low accuracy
- Accurate VOP detection may improve the CV recognition performance
- The recognition performance of CV units by using single level hybrid approach presented in [58] can be improved by using multi-level hybrid approach

An approach based on combination of SVM and HMM evidences for enhancing the CV recognition performance is presented in [58]. A summary of the discussion related to CV recognition is given in Table 2.3.

2.4   Time Scale Modification

The purpose of time scale modification (TSM) is to change the rate of speech while preserving the characteristics of the original speech such as formant structure and pitch periods. There are various applications of time scale modification. For example, time scale compression can be used at the input side of the speech coder and transmission, followed by time scale expansion at the receiver to get back the original speech [59]. In some other applications, time scale expansion can be used to enhance the intelligibility of rapid or degraded speech [60]. Time scale compression is also useful in message playback systems for fast scanning of recorded messages [61]. Recently, adaptive TSM is used in VoIP applications for handling the network congestion [62]. Pitch and time scale modification was attempted in real time by focusing on processing only voiced regions of speech utterance [63].

There are number of approaches available in the literature for time scale modification. Some of them use sinusoidal model, pitch synchronous overlap and add (PSOLA), and phase vocoders [60, 64, 65]. In [59], the authors have presented an epoch-based time scale modification method, where the duration of speech signal is modified using the knowledge of epochs. In this method, TSM is performed in residual domain. Linear prediction pitch synchronous overlap and add (LP-PSOLA) approach also performs TSM in the residual domain similar to the epoch-based method [66]. The approaches mentioned above basically perform time scale modification uniformly, for entire speech signal. But, fast or slow speech produced by humans may not vary uniformly across all the speech segments. In [16], fast and slow speech produced by human beings was analyzed, and observed that durations of consonant and transition regions remain the same in fast or slow speech, and only the vowel and pause regions will vary according to speech rate. Based on this observation, authors have presented a nonuniform time scale modification method, where consonant and transition regions of speech are kept unaltered, and only vowel and pause segments are modified according to desired speaking rate [16].

Table 2.4 Summary of the review of TSM methods

- Majority of the existing TSM methods modify all speech segments with same modification factors
- Modifying different speech segments with different modification factors based on their production and articulatory constraints may improve the quality of speech
---

Attempts to incorporate nonuniform duration modification are reported in the literature [64, 67, 68]. Speech adaptive TSM method presented in [64] modify the speech rate based on voicing probability derived from sinusoidal pitch estimator. The voicing probability is close to unity during steady voicing, decreases during transition, and close to zero during unvoiced speech and pauses. The assumption is that changes in speaking rate for compression or expansion do not take place in sounds which are not voiced, but they occur mostly in voiced sounds. A nonuniform time scaling method has been developed along with spectral shape and pitch modification for automatic morphing of one sound to another sound [67]. Another method for speech adaptive TSM is presented, which allows slowing down the speech without compromising the quality or naturalness of the slowed speech [68]. In this method, different scaling factors are applied to different types of speech segments. Transient detection in music and audio signals has been studied for different applications such as segmentation and editing of audio recordings [69] and improving audio effects [70] through TSM [71–73]. Different methods use different cues of audio signal for the detection of transient audio segments. Sum of significant spectral peaks is used in [74] for discriminating the transients from steady segments. Variance of the spectrum and time offset of the center of gravity are used in [75] for classifying the transients. In most of the studies, transient detection was used to improve the quality of audio for different speaking rates. Bonada has proposed a frequency domain method for processing the fast changes in the signal in a different way compared to other components [71]. Roebel has proposed a new approach for processing transients in the phase vocoder, where transient peaks are preserved during stretching [73]. Recently, a nonuniform TSM method based on waveform similarity overlap-and-add (WSOLA) technique is presented for time scale modification of music signals [76]. In this approach, the perceptually significant transient sections (PSTs) such as temporal envelope changes and significant spectral transitions will be preserved from modification. A summary of existing approaches for TSM is given in Table 2.4.

2.5 Summary

In this chapter, we have reviewed some of the existing methods for VOP detection, speech systems in mobile environment, CV recognition in Indian languages, and time scale modification. Existing methods for VOP detection are suffering with poor detection accuracy. Therefore, accuracy issues in the detection of VOP are the main focus in this work. In contrast to the existing block processing approaches, the methods proposed in this work enhance VOP detection performance by exploiting the spectral energy in glottal closure regions. The goal of this book is to demonstrate the significance of accurate VOP detection for CV recognition, speaker identification, and nonuniform time scale modification.

Chapter 3
Vowel Onset Point Detection from Coded and Noisy Speech

Abstract Most of the existing vowel onset point (VOP) detection methods are developed for clean speech. In this chapter, we propose methods for detection of VOPs in the presence of speech coding and background noise conditions. VOP detection method for coded speech is based on the spectral energy between 500 and 2,500 Hz frequency band of the speech segments present in glottal closure region. In case of noisy speech, the proposed VOP detection method exploits the spectral energy at the formant locations of the speech segments present in glottal closure region. The proposed VOP detection methods are evaluated using objective measures and consonant vowel (CV) unit recognition experiments.

This chapter discusses about the recently proposed robust vowel onset point (VOP) detection methods in the presence of speech coding and background noisy environments. The proposed VOP detection methods are thoroughly evaluated using isolated consonant vowel utterances and spotting CV units from continuous speech. This chapter is organized as follows: Speech databases used for the evaluation of VOP detection methods are described in Sect. 3.1. Section 3.2 discusses the VOP detection method for coded speech. Performance of the VOP detection methods in the presence of coding is presented in Sect. 3.3. Section 3.4 presents the VOP detection method for noisy speech. In Sect. 3.5, performance of the VOP detection methods is evaluated in the presence of noisy conditions. Finally Sect. 3.6 summarizes the performance of VOP detection methods in the presence of coding and noise.

3.1 Speech Databases for VOP Detection

In this work, VOP detection methods are evaluated using TIMIT database and Telugu broadcast news database. Details of the databases are briefly described in the following subsections.

K.S. Rao and A.K. Vuppala, *Speech Processing in Mobile Environments*, SpringerBriefs in Electrical and Computer Engineering, DOI 10.1007/978-3-319-03116-3_3,

3.1.1 TIMIT Database

The TIMIT database is collected from American English speakers, divided into
eight accent regions including speakers who do not have strong regional accents
[77]. The database contains data from 630 speakers, of which 438 are males (70%)
and 192 are females (30%). The text material in the TIMIT prompts (found in the
file "prompts.doc") consists of phonetically compact sentences designed at MIT and
phonetically diverse sentences selected at TI. The phonetically compact sentences
were designed to provide a good coverage of pairs of phones, with extra occurrences
of phonetic contexts thought to be either difficult or of particular interest. The
phonetically diverse sentences were selected to add diversity in sentence types and
phonetic contexts. Each speaker has contributed ten sentences of approximately 3 s
each. The speech was recorded using a high quality microphone in sound proof
booth with no session interval between recordings. The TIMIT database contains
speech files with manually marked phoneme boundaries, and these phoneme
boundaries are used for marking the reference VOPs. The VOPs hypothesized by
an automatic method may be compared with reference VOPs to find the deviations,
spurious VOPs, and missed VOPs.

3.1.2 Broadcast News Database

Broadcast news database was developed at speech and vision lab, Indian Institute
of Technology, Madras, India [16, 54–56, 78–82]. This database has been extensively used for developing various speech systems and analyzing the speech data
for various speech applications in the context of Indian languages [54, 83–86].
Broadcast news corpus is recorded for four Indian languages, namely Telugu, Tamil,
Kannada, and Hindi. Among four languages, we consider Telugu language news
database for this work. Duration of Telugu broadcast database is about 5 h, collected
over 20 bulletins by 11 male speakers and 9 female speakers. Database is transcribed
manually at phrase, word, and syllable levels. Manually marked syllable boundaries
are used for picking the CV units from continuous speech utterances. In this work,
95 CV classes whose frequency of occurrence in the database is more than 50 are
considered for the analysis, and their contribution is more than 95% of CV units
present in the database.

3.2 VOP Detection Method for Coded Speech

The major motivation for the proposed VOP detection method is the retention of
spectral characteristics of speech signals by the low-bit rate speech coders. Proposed
VOP detection method uses the spectral energies of speech segments present in the

glottal closure regions of voiced speech [87]. In the existing spectral energy-based VOP detection method, spectrum is derived using conventional block processing with a frame size of 20 ms and frame shift of 10 ms [15]. In general, we assume that speech signal in voiced region is stationary within 20–30 ms, but there exists a non-stationary behavior even in between two consecutive pitch cycles [6, 88, 89]. Therefore, spectrum estimated from 20 ms frame corresponds to the average spectral characteristics of multiple pitch cycles present in the frame.

A pitch cycle (glottal cycle) is a combination of glottal closure and glottal open phases. During glottal closure phase, vocal tract is completely isolated from trachea and lungs. Spectrum estimation during glottal closure phase will be more accurate because true vocal tract (oral cavity) resonances will be present during that phase, whereas in glottal open phase the spectrum refers to the combination of oral cavity, trachea, and lungs cavity. This is due to coupling of oral cavity with trachea and lungs during open phase of vocal folds. Therefore, the spectrum derived from the block processing consists of a mixture of vocal tract resonances and resonances due to oral, trachea, and lung cavities together. In the present work, for detecting the VOPs, spectral energy at the glottal closure region is used as evidence. Glottal closure instant (GCI) indicates starting of glottal closure phase. Therefore, speech segment considered in this work for estimating the spectrum is 30% of glottal cycle (pitch period) starting from the glottal closure instant. The reason for choosing 30% of glottal cycle is to ensure that the chosen speech segment should generate during glottal closure phase. It is also known that speech signal during glottal closure phase has high signal-to-noise ratio compared to other regions. Therefore, the spectral energy in glottal closure region is high compared to glottal open region [6, 88]. To identify the glottal closure region, we use glottal closure instant as the beginning of glottal closure phase.

The glottal closure instants are also known as instants of significant excitation or epochs. Among various existing methods to determine the epoch locations [90–92], in this work zero frequency filter (ZFF) method has been used for estimating the epoch locations [92]. ZFF method will give accurate GCI locations during voiced speech and random locations for unvoiced speech. ZFF method for epoch extraction provides 99% of accuracy for clean speech and its performance is also robust to noisy conditions [92]. The effect of speech coding on epoch extraction is reported in [93]. The performance of ZFF method is superior among existing methods, and hence it is used in this work for extracting epoch locations under both clean and degraded conditions. In Sect. 3.2.1, sequence of steps for the extraction of glottal closure instants (epoch locations) using ZFF method is described. Sequence of steps in the proposed VOP detection method is presented in Sect. 3.2.2. The choice of frame size used in the proposed method is analyzed in Sect. 3.2.3. The choice of 500–2,500 Hz frequency band for deriving the spectral energy in the proposed VOP detection method is justified in Sect. 3.2.4.

3.2.1　Extraction of Glottal Closure Instants Using ZFF Method

Among the existing epoch extraction methods, ZFF method determines the epoch locations with highest accuracy [92]. ZFF exploits the discontinuities due to impulse excitation reflected across all the frequencies including the zero frequency. The influence of vocal tract system is negligible at zero frequency. Therefore, zero frequency filtered speech signal carries excitation source information, which is used for extracting the epoch locations. The ZFF method consists of the following sequence of steps:

- Difference the input speech signal to remove any time-varying low frequency bias in the signal

$$x(n) = s(n) - s(n-1) \tag{3.1}$$

- Compute the output of cascade of two ideal digital resonators at 0 Hz, i.e.,

$$y(n) = \sum_{k=1}^{4} a_k y(n-k) + x(n) \tag{3.2}$$

where $a_1 = +4$, $a_2 = -6$, $a_3 = +4$, $a_4 = -1$. Note that this is equivalent to passing the signal $x(n)$ through a digital filter given by

$$H(z) = \frac{1}{(1 - z^{-1})^4} \tag{3.3}$$

- Remove the trend, i.e.,

$$\hat{y}(n) = y(n) - \bar{y}(n) \tag{3.4}$$

where

$$\bar{y}(n) = \frac{1}{2N+1} \sum_{n=-N}^{N} y(n) \tag{3.5}$$

Here $2N+1$ corresponds to the size of the window used for computing the local mean, which is typically the average pitch period computed over a long segment of speech.
- The trend removed signal $\hat{y}(n)$ is termed as *zero frequency filtered* (ZFF) signal. Positive zero-crossings in the ZFF signal correspond to the epoch locations.

Epoch extraction for the segment of voiced speech using ZFF method is shown in Fig. 3.1. Figure 3.1a shows the differenced electro-glottograph (EGG) signal

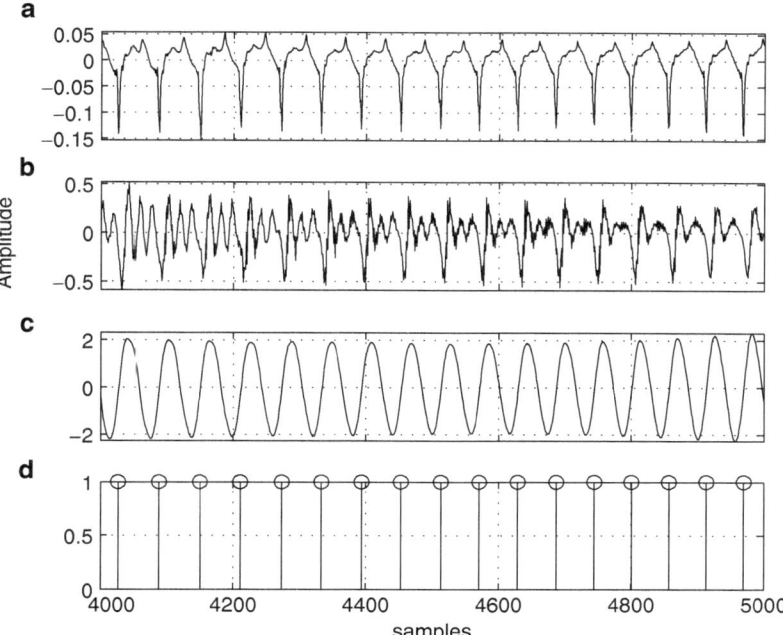

Fig. 3.1 Epoch (GCI) extraction using zero frequency filtering method. (**a**) Differenced EGG signal, (**b**) speech signal, (**c**) zero frequency filtering signal, and (**d**) epochs (GCIs) derived from zero frequency filtered signal

of voiced speech segment shown in Fig. 3.1b. ZFF signal and the derived epoch locations are shown in Fig. 3.1c, d, respectively. From Fig. 3.1a, d, it is evident that the epochs extracted using ZFF method almost coincide with the negative peaks of differenced EGG signal, which indicate the instants of glottal closure.

3.2.2 Sequence of Steps in the Proposed VOP Detection Method

1. Determine the epoch locations (glottal closure instants) by using ZFF method.
2. Compute discrete Fourier transform (DFT) for the window (W) of speech samples present in 30% of glottal cycle starting from the GCI. Window W is computed by using below equations:

$$GC = [gci(i + 1) - gci(i)] \qquad (3.6)$$

Where GC is glottal cycle, $gci(i + 1)$ and $gci(i)$ are glottal closure instants at i and $i + 1$ locations.

$$WL = int(0.30 \times GC) \tag{3.7}$$

where WL is window length.

$$W = [s(gci(i)), s(gci(i) + 1), \ldots, s(gci(i) + WL - 1)] \tag{3.8}$$

where W is window and s is speech samples.

3. Determine the spectral energy within the frequency band of 500–2,500 Hz. Here spectral energy in 500–2,500 Hz band is considered, where energy of the vowel is much higher than the consonant.
4. Spectral energy is plotted as a function of time. Fluctuations in the spectral energy contour are smoothed by using mean smoothing with 50 ms window.
5. The change at the VOP present in the smoothed spectral energy of the speech signal is enhanced by computing its slope using first-order difference (FOD). FOD of $x(n)$ is given by

$$x_d(n) = x(n) - x(n-1) \tag{3.9}$$

The finer details involved in the enhancement of VOP evidence are illustrated by using Fig. 3.2. Figure 3.2a shows the speech utterance. Smoothed spectral energy in 500–2,500 Hz band around each epoch is shown in Fig. 3.2b. The FOD signal of smoothed spectral energy is shown in Fig. 3.2c. Since FOD values corresponding to slopes, positive to negative zero crossings of slopes correspond to local peaks in the smoothed spectral energy signal. These local peaks are shown by star (*) symbols in Fig. 3.2b. The unwanted peaks in Fig. 3.2b are eliminated by using the sum of slope values within 10 ms window centered at each peak. Figure 3.2d shows the sum of slope values within 10 ms around each peak. The peaks with the lower sum of slope values are eliminated with a threshold set to 0.5 times the mean value of the sum of slopes. This threshold is determined empirically. Further, if two successive peaks present within 50 ms, then the lower peak among the two will be eliminated, based on the assumption that two VOPs won't present within 50 ms interval. The desired peak locations are shown in Fig. 3.2e with star (*) symbol after eliminating the unwanted peaks. At each local peak location, the nearest negative to positive zero crossing points (see Fig. 3.2c) on either side are identified and marked by circles on Fig. 3.2e. The regions bounded by negative to positive zero crossing points are enhanced by normalization process shown in Fig. 3.2f. Here, normalized values are computed by using the below equation:

$$N(i) = \frac{x(i) - min}{max - min} \tag{3.10}$$

where $N(i)$ is normalized value of input x(i); min and max are local minimum and maximum.

6. Significant changes in spectral characteristics present in the enhanced version of the smoothed spectral energy are detected by convolving with first order

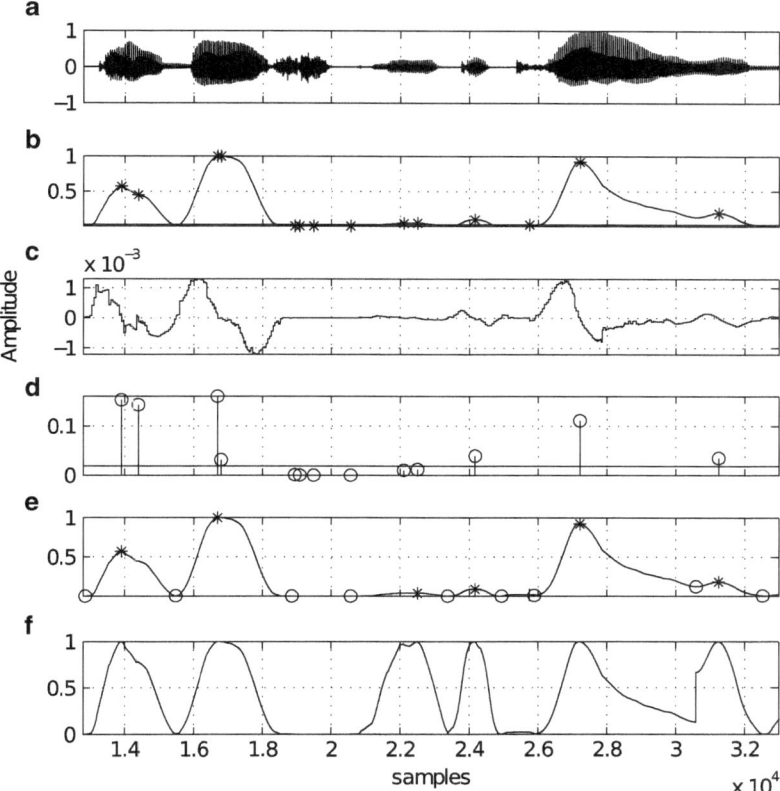

Fig. 3.2 Enhancement of VOP evidence for a speech utterance /"*Don't ask me to carry an*"/. (**a**) Speech signal, (**b**) smoothed spectral energy in 500–2,500 Hz band around each epoch, (**c**) FOD values, (**d**) sum of slope values computed at each peak locations, (**e**) smoothed spectral energy plot with peak locations, and (**f**) enhanced values

Gaussian difference (FOGD) operator of length 100 ms. A Gaussian window $g(n)$ of length L is given by

$$g(n) = \frac{1}{\sqrt{2\pi}\sigma} e^{-\frac{n^2}{2\sigma^2}}, \qquad n = 1, 2, \ldots, L \qquad (3.11)$$

where σ is standard deviation. In this work, σ value of 200 is considered. The choice of length of Gaussian window (L) is based on an assumption that the VOP occurs as gross level changes at intervals of about 100 ms [12, 15]. FOGD is given by $g_d(n)$, and it is shown in Fig. 3.3.

$$g_d(n) = g(n) - g(n-1) \qquad (3.12)$$

The convolved output is the proposed VOP evidence plot.

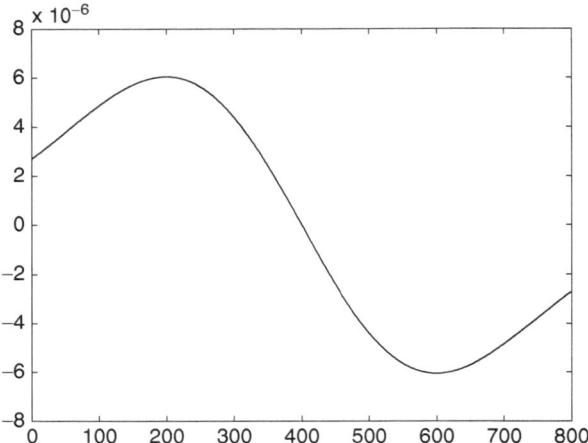

Fig. 3.3 FOGD operator with $L = 800$, and $\sigma = 200$

7. Positive peaks in the VOP evidence plot represent the VOP locations. The flow diagram of the proposed VOP detection method is shown in Fig. 3.4.

Output of each step in the proposed method is shown in Fig. 3.5 by using speech utterance /"*Don't ask me to carry an*"/. Figure 3.5a shows the speech signal with manually marked VOPs. Spectral energy in 500–2,500 Hz band and its smoothed signal are shown in Fig. 3.5b, c respectively. Figure 3.5d shows the enhanced signal correspond to the signal present in Fig. 3.5c. Figure 3.5e shows the VOP evidence signal obtained by convolving the enhanced spectral energy signal with FOGD. We can observe that manual VOPs marked in Fig. 3.5a and detected VOPs marked in Fig. 3.5e are close to each other.

VOP detection using proposed and existing methods is shown in Fig. 3.6. Figure 3.6a shows the speech segment for the utterance /"*Don't ask me to carry an*"/. VOP evidence plots for the speech signal shown in Fig. 3.6a using excitation source, spectral peaks, modulation spectrum, COMB-ESM, and proposed methods are shown in Fig. 3.6b–f respectively. In Fig. 3.6 it is observed that spurious VOP (third one) present in COMB-ESM and individual methods is eliminated in proposed method.

3.2.3 Choice of Frame Size

In the proposed VOP detection method, the size of speech frame to be considered at each epoch should fall within the glottal closure interval. But determining the glottal closure region precisely within each glottal cycle is difficult. Therefore, we have analyzed various frame durations varying from 10% to 60% of pitch period

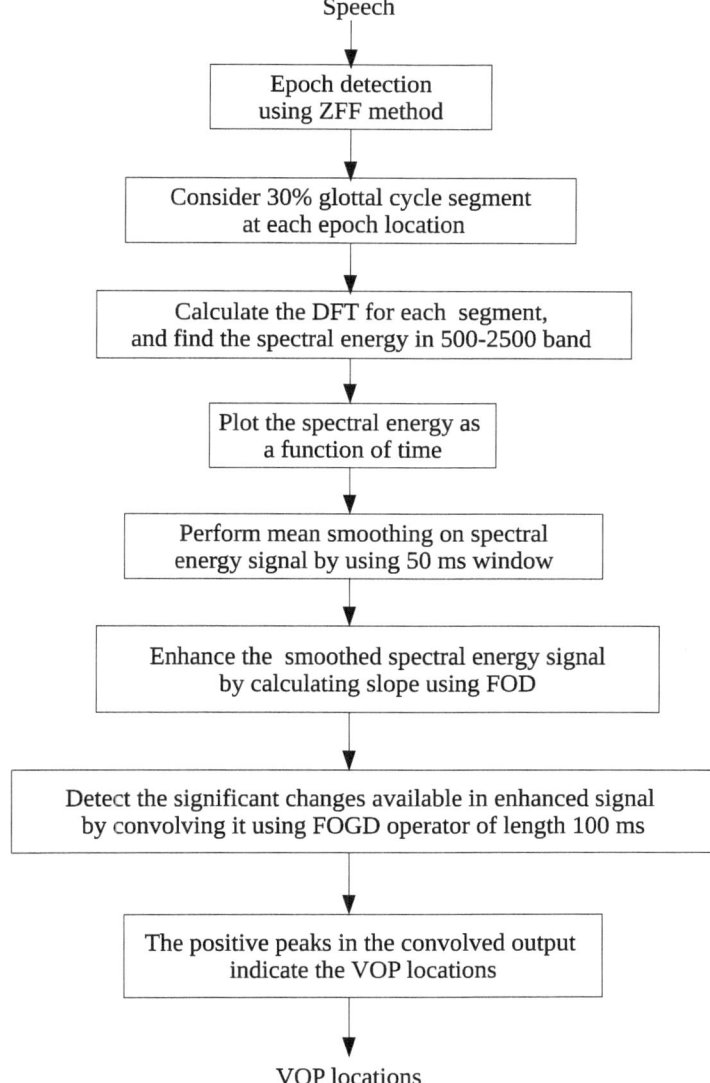

Speech

Epoch detection
using ZFF method

Consider 30% glottal cycle segment
at each epoch location

Calculate the DFT for each segment,
and find the spectral energy in 500-2500 band

Plot the spectral energy as
a function of time

Perform mean smoothing on spectral
energy signal by using 50 ms window

Enhance the smoothed spectral energy signal
by calculating slope using FOD

Detect the significant changes available in enhanced signal
by convolving it using FOGD operator of length 100 ms

The positive peaks in the convolved output
indicate the VOP locations

VOP locations

Fig. 3.4 Flow diagram of the proposed VOP detection method for coded speech

for choosing the appropriate frame size to represent the glottal closure region. For analyzing the effect of frame size on VOP detection, 110 sentences from TIMIT database are considered. Table 3.1 shows the performance of VOP detection using proposed method by considering different durations of speech frames in the glottal cycle. Column-1 indicates different durations of glottal closure speech segments considered for calculating the spectral energy. Column-2 indicates the

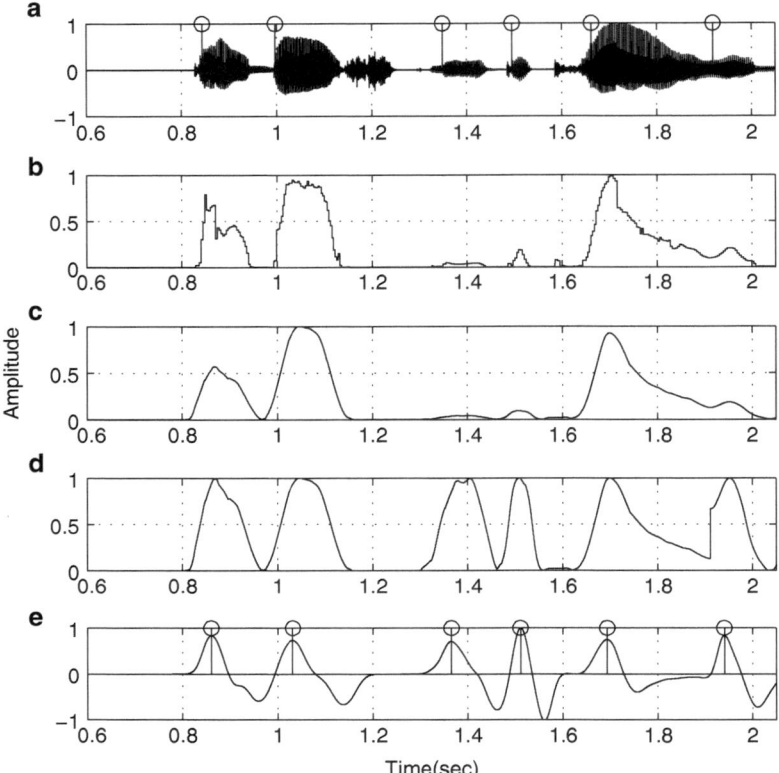

Fig. 3.5 VOP detection using proposed method for a speech utterance /"*Don't ask me to carry an*"/. (**a**) Speech signal with manually marked VOPs, (**b**) spectral energy in 500–2,500 Hz band around each epoch, (**c**) mean smoothed spectral energy, (**d**) enhanced spectral energy signal, and (**e**) proposed VOP evidence signal

percentage of VOPs detected within 40 ms deviation. In this study 40 ms is the maximum deviation considered between reference and detected VOPs. Column-3 indicates the percentage of missed VOPs.

From the results presented in Table 3.1, it is observed that accuracy in the detection of VOPs is optimal by using 30% of glottal cycle as frame size in the proposed method compared to other segment durations. Hence, 30% of glottal cycle is considered in the proposed method for determining the VOPs. Robustness of the proposed VOP detection method is analyzed by using speech utterances with different average pitch periods. For performing this study, we considered three sets of utterances with pitch periods varying from 2.5–5, 5–7, and 7–10 ms. Thirty utterances are recorded from children (in age group of 6–10 years) to cover 2.5–5 ms pitch periods. Utterances having 5–7 and 7–10 ms pitch periods are taken from TIMIT database, and each set contains 30 utterances. In this study, 30% of

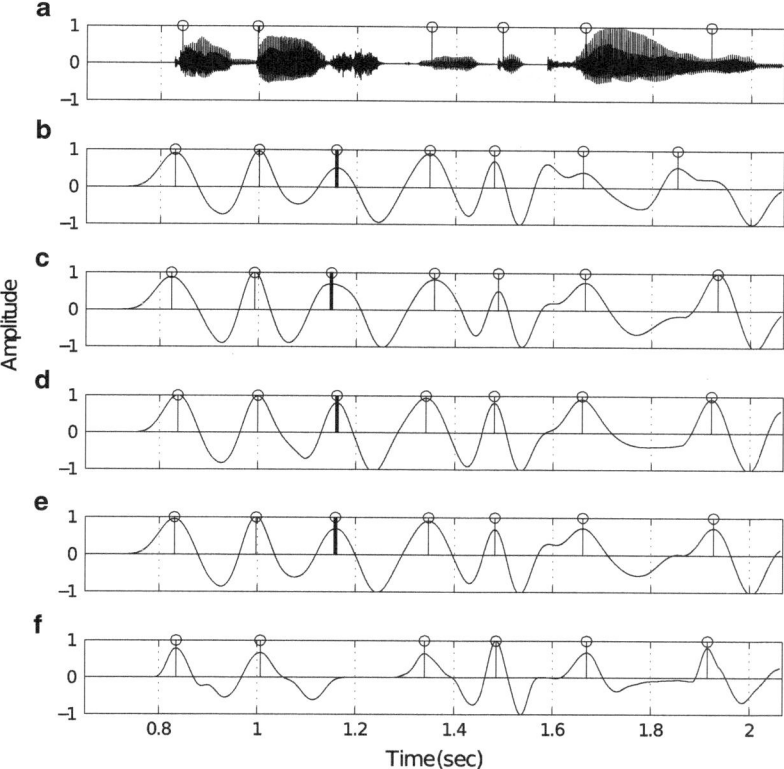

Fig. 3.6 VOP detection using existing and proposed VOP detection methods for a speech utterance /*"Don't ask me to carry an"*/. (**a**) Speech signal, VOP evidence plots for (**b**) excitation source, (**c**) spectral peaks, (**d**) modulation spectrum, (**e**) COMB-ESM method, and (**f**) proposed method

pitch period is used for processing the speech segments in glottal closure region. Table 3.2 shows the performance of VOP detection for speech utterances having different average pitch periods. Column-1 indicates range of average pitch periods, and column-2 indicates corresponding F0 values. Column-3 indicates the percentage of VOPs detected within the 40 ms deviation. Column-4 indicates the percentage of missed VOPs. From the results, it is observed that proposed VOP detection method is robust to speech signals with different pitch periods.

3.2.4 Choice of Frequency Band

In this work, VOP evidence is derived by computing spectral energy of speech segment at each epoch. If the epoch is happened to be GCI in the vowel region,

Table 3.1 Performance of
the proposed VOP detection
method for different frame
sizes considered at GCIs

Percentage of pitch period	VOP detection (%) with 40 ms deviation	Missing rate
10	93.61	6.39
20	93.94	6.06
30	**95.3**	**4.5**
40	94.52	5.48
50	94.16	5.84
60	93.92	6.08

Table 3.2 Performance of the proposed VOP detection method for speech utterances with different average pitch periods

Avg. pitch period (ms)	F0 values (Hz)	VOP detection (%) with 40 ms deviation	Missing (%) rate
2.5–5	201–400	95.21	4.79
5–7	143–200	95.44	4.56
7–10	100–143	95.31	4.69

the spectral energy within 500–2,500 Hz band is very high compared to consonant and nonspeech regions. For voiced consonants or nasals most of the spectral energy is present below 500 Hz. For unvoiced consonants, fricatives, and other sound units, most of the spectral energy is present beyond 3,000 Hz. Therefore, the spectral energy in frequency band 500–2,500 Hz will provide accurate VOP evidence compared to sum of 10 spectral peaks used in the existing methods.

VOP detection using proposed method by considering spectral energy in different frequency bands for a speech utterance /"*Don't ask me to carry an*"/ is shown in Fig. 3.7. Figure 3.7a shows the speech signal. VOP evidence plots using proposed method by considering spectral energy in 0–4,000 Hz, 750–2,500 Hz, 500–2,500 Hz, 1,000–4,000 Hz, and 2,500–4,000 Hz bands are shown in Fig. 3.7b–f, respectively. In Fig. 3.7c, it is observed that third VOP is missed. In Fig. 3.7e, f, spurious VOPs can be observed (VOPs indicated in bold). In Fig. 3.7b, d only genuine VOPs are observed. Choice of 500–2,500 Hz band in the proposed method is extensively studied by considering different frequency bands. Table 3.3 shows the performance of VOP detection for consonant-vowel (CV) units from Telugu broadcast news speech corpus using proposed method by considering spectral energy at different frequency bands. Various frequency bands are considered in 0–4,000 Hz range. For evaluation, 950 (10 utterances from each 95 most frequently occurred CV units) CV utterances are considered to cover various consonant and vowel combinations. Therefore, this evaluation also ensures the robustness of the chosen frequency band in the proposed VOP detection method for different classes of CV speech segments. Column-1 indicates different frequency bands considered for calculating the spectral energy. Column-2 indicates the percentage of VOPs detected within the 40 ms deviation. Column-3 indicates the percentage of missed VOPs. Column-4 indicates the average deviation (in ms) with respect to the manual marked VOPs. From Table 3.3, it is observed that accuracy in the detection of VOPs is better

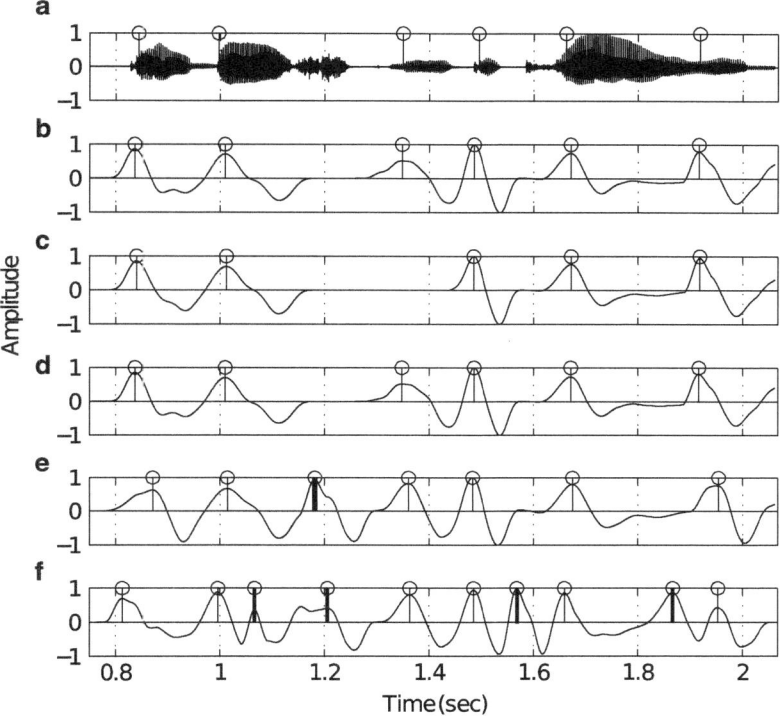

Fig. 3.7 VOP detection using proposed method for different frequency bands on a speech utterance /*"Don't ask me to carry an"*/. (**a**) Speech signal, VOP evidence plot by considering spectral energy from (**b**) 0–4,000 Hz, (**c**) 750–2,500 Hz, (**d**) 500–2,500 Hz, (**e**) 1,000–4,000 Hz, and (**f**) 2,500–4,000 Hz

Table 3.3 Performance of the proposed VOP detection method for consonant-vowel (CV) units from Telugu broadcast news speech corpus by considering spectral energy at different frequency bands

Frequency band	VOP detection (%) with 40 ms deviation	Missing rate (%)	Average deviation (ms)
0–4,000	95.6	4.4	12.87
0–2,500	96.5	3.5	13.23
0–2,000	96.56	3.44	13.23
250–4,000	96.56	3.44	12.67
250–3,750	96.11	3.89	12.87
250–3,250	96.56	3.44	12.69
250–2,500	97.19	2.81	12.99
500–2,750	96.88	3.12	12.25
500–2,500	**97.5**	**2.5**	**11.22**
500–4,000	96.88	3.12	12.88
750–2,500	93.44	6.56	14.70

by using 500–2,500 Hz frequency band in the proposed method compared to other bands. Hence, in our further studies 500–2,500 Hz frequency band is used in the proposed method for determining the VOPs.

3.3 Performance of the VOP Detection Method in the Presence of Speech Coding

In this section performance of the proposed VOP detection method is compared with COMB-ESM method (described in Sect. 2.1) which uses the combination of evidences from excitation source (EXC), spectral peaks (SP), and modulation spectrum (MOD). VOP detection methods are evaluated on continuous speech and CV units from broadcast speech. Speech coders considered in this work are GSM full rate (ETSI 06.10), GSM enhanced full rate (ETSI 06.60), CELP (FS-1016), and MELP (TI 2.4 kbps). GSM full rate coder provides 13 kbps bit rate using regular pulse excitation and long-term prediction (RPE-LTP) technique. GSM enhanced full rate (EFR) coder is designed to improve GSM full rate coder under channel error conditions. GSM EFR coder provides 12.2 kbps bit rate by using algebraic code excited linear prediction (ACELP) scheme. CELP coder provides 4.8 kbps bit rate using code excited linear prediction technique. MELP coder provides 2.4 kbps bit rate using mixed excitation linear prediction technique. Brief details about the speech coders considered in this work are given in Appendix B.

3.3.1 VOP Detection from Continuous Speech Under Coding

VOP detection studies are conducted on TIMIT database [77] for analyzing the performance of VOP detection methods. Coded speech data is prepared by passing clean speech data through encoder and decoder of the standard coders. GSM full rate (FR), GSM enhanced full rate (EFR), CELP, and MELP speech coders are considered in this study. About 110 sentences (60 sentences spoken by female speakers and 50 sentences are spoken by male speakers) having 1,197 manually marked VOPs are considered for analyzing the performance of the proposed VOP detection method. Among 1,197 VOPs, 534 VOPs correspond to the utterances spoken by male speakers, and the rest 663 VOPs correspond to the utterances spoken by female speakers.

Performance of different VOP detection methods is compared using parameters like average deviation, missing rate, and spurious rate. VOPs detected within 40 ms deviation to the reference VOPs are considered as genuine VOPs. The ratio (in %) of number of genuine VOPs detected to the total number of reference VOPs is measured for different time resolutions (10–40 ms). Average deviation (in ms) is calculated from the deviations of genuine detected VOPs. The ratio (in %) of

Table 3.4 Performance of the VOP detection using excitation source (EXC), spectral peaks (SP), modulation spectrum (MOD), COMB-ESM, and proposed methods on TIMIT database (1,197 reference VOPs) in the presence of coding

VOP detection method	Hypothe-sized VOPs	VOPs detected within ms ($\approx$ %)				AVG dev. ($\approx$ ms)	MISS VOPs ($\approx$ %)	SPU VOPs ($\approx$ %)
		10	20	30	40			
Clean								
EXC	1,176	34	49	59	94	20	6	4
SP	1,172	29	48	70	93	21	7	5
MOD	1,126	33	50	73	92	18	8	2
COMB-ESM	1,173	51	59	70	95	16	5	3
Proposed	**1,162**	**65**	**83**	**91**	**95**	**12**	**5**	**2**
GSM FR								
EXC	1,162	31	47	57	91	21	9	6
SP	1,159	28	47	68	92	21	8	5
MOD	1,066	27	45	65	85	20	15	4
COMB-ESM	1,114	45	51	65	90	18	10	3
Proposed	**1,164**	**63**	**81**	**88**	**94**	**13**	**6**	**3**
GSM EFR								
EXC	1,163	32	48	57	92	20	8	5
SP	1,160	28	46	69	92	20	8	5
MOD	1,088	29	47	67	88	19	12	3
COMB-ESM	1,136	47	55	68	92	17	8	3
Proposed	**1,152**	**63**	**82**	**89**	**94**	**13**	**6**	**2**
CELP								
EXC	1,103	25	40	50	84	23	16	8
SP	1,114	26	42	64	87	22	13	6
MOD	1,089	24	41	62	85	21	15	6
COMB-ESM	1,102	34	46	59	86	20	14	6
Proposed	**1,126**	**39**	**67**	**80**	**88**	**17**	**12**	**6**
MELP								
EXC	1,112	26	42	52	86	22	14	7
SP	1,125	28	46	65	89	21	11	5
MOD	1,054	26	44	64	84	20	16	4
COMB-ESM	1,090	38	48	59	87	19	13	4
Proposed	**1,138**	**52**	**72**	**84**	**91**	**15**	**9**	**4**

undetected VOPs to the total number of reference VOPs is termed as missing rate (MISS). VOPs detected other than genuine VOPs are termed as spurious VOPs. The ratio of spurious VOPs (in %) detected to the number of reference VOPs is termed as spurious rate (SPU).

Table 3.4 shows the accuracy in the detection of VOPs using different methods in the presence of various coders. Column-1 indicates different methods considered in the analysis for detecting the VOPs. Column-2 indicates the total number of VOPs detected by various methods. Columns 3–6 indicate the percentage of

VOPs detected within the specified deviations. Column-7 indicates the average deviation (in ms) with respect to the manual marked VOPs. Columns 8 and 9 indicate the percentage of missed and spurious VOPs, respectively. From the results, it is observed that coding has significant effect on VOP detection performance. Spurious VOPs are observed to be increased due to coding. Accuracy in detection of VOPs is observed to be superior using the proposed method under both clean and coded cases, compared to existing methods. Average deviation has reduced significantly, in case of proposed method compared to other methods. Average deviation using proposed method is around 1 ms, 1 ms, 5 ms, and 3 ms higher compared to clean case for GSM FR, GSM EFR, CELP, and MELP coders, respectively.

VOP detection from CELP coded speech is observed to be less accurate compared to MELP coder even though the bit rate provided by MELP coder is less than CELP coder. The reason may be due to poor way of representation of excitation signal in CELP coding technique. CELP coder uses the code book to represent the excitation signal, which introduce more approximation compared to other coders. From the results, it is observed that performance of VOP detection methods based on the spectral energy is superior compared to the other methods in the presence of coding. Among the methods based on spectral energy, performance of the proposed method is better in all aspects. The improved performance of the proposed method is due to exploiting the high SNR characteristics of speech signal present in the glottal closure phase.

3.3.2 VOP Detection from CV Units Under Coding

CV units collected from Telugu broadcast news corpus are used for evaluation. In this study, 95 CV classes are considered. From each CV class ten utterances are considered, with this total number of VOPs used in this study is 950. In this study, analysis of spurious VOPs is not applicable, since each utterance contains only one VOP, and it is determined based on the peak in the VOP evidence signal. VOP detection using proposed and COMB-ESM methods under different coding cases, for a speech utterance $/ba/$ is shown in Fig. 3.8. Figure 3.8a, d, g, j shows the clean, GSM, CELP, and MELP-coded speech segments, respectively. Figure 3.8(b, c),(e, f),(h, i), and (k, l) shows the COMB-ESM and proposed VOP evidence plots for clean, GSM, CELP and MELP speech segments, respectively. From the Fig. 3.8, it is observed that VOP detection accuracy using proposed method is superior compared to COMB-ESM method.

Table 3.5 shows the accuracy of VOP detection for Telugu broadcast news data using the proposed and existing methods. Column-1 indicates different methods considered in the analysis for detecting the VOPs. Columns 2–5 indicate the percentage of VOPs detected within the specified deviations. Column-6 indicates the average deviation with respect to the manual marked VOPs. Column-7 indicates the percentage of missed VOPs. From Table 3.5, it is observed that accuracy in

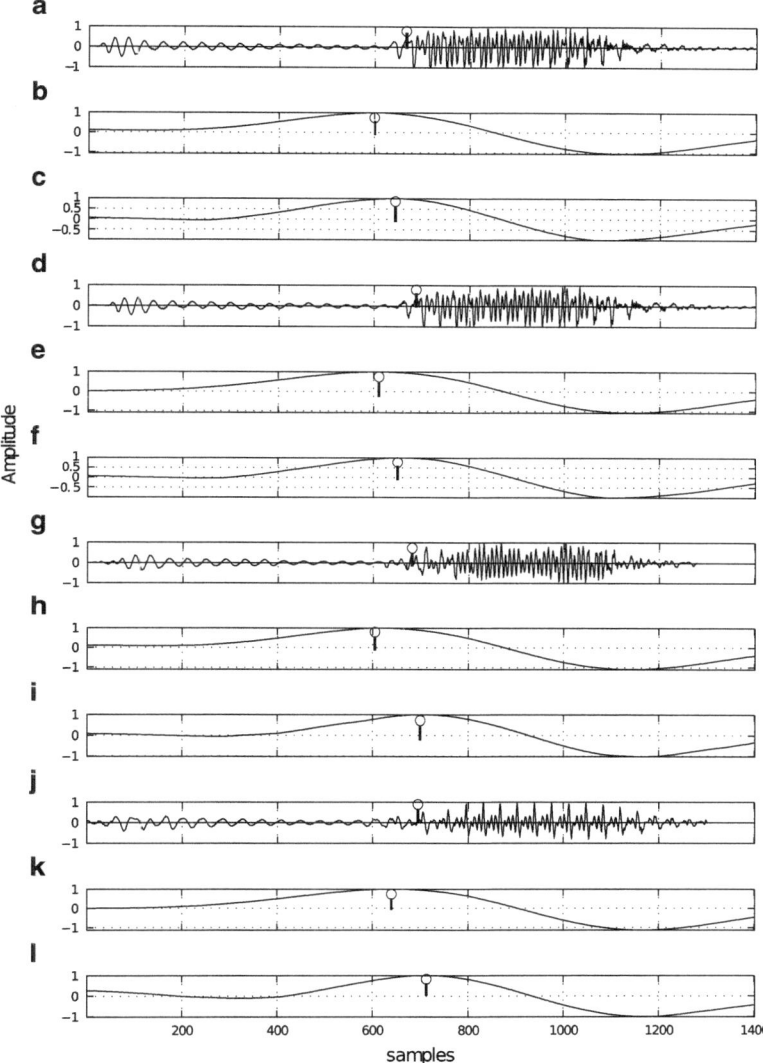

Fig. 3.8 VOP detection using COMB-ESM and proposed methods for utterance /ba/. (**a**) Speech signal. (**b**) and (**c**) are VOP evidence plots using COMB-ESM and proposed method, respectively, for the speech signal shown in (**a**). (**d**) GSM-FR coded speech. (**e**) and (**f**) are VOP evidence plots using COMB-ESM and proposed method, respectively, for the GSM-FR-coded speech signal shown in (**d**). (**g**) CELP-coded speech. (**h**) and (**i**) are VOP evidence plots using COMB-ESM and proposed method, respectively, for the CELP-coded speech signal shown in (**g**). (**j**) MELP-coded speech. (**k**) and (**l**) are VOP evidence plots using COMB-ESM and proposed method, respectively, for the MELP-coded speech signal shown in (**j**)

Table 3.5 Performance of the VOP detection using excitation source (EXC), spectral peaks (SP), modulation spectrum (MOD), and COMB-ESM (COMB) methods on CV units from broadcast news speech corpus (950 VOPs are considered) in the presence of coding

VOP detection method	VOPs detected within ms ($\approx$ %)				AVG dev. ($\approx$ ms)	MISS VOPs ($\approx$ %)
	10	20	30	40		
Clean						
EXC	44	69	84	90	17	10
SP	48	78	92	95	13	5
MOD	36	50	74	92	20	8
COMB-ESM	50	74	89	95	15	5
Proposed	**69**	**88**	**93**	**97**	**11**	**3**
GSM FR						
EXC	42	67	82	89	18	11
SP	46	75	90	94	14	6
MOD	32	44	67	87	21	13
COMB-ESM	45	72	86	91	17	9
Proposed	**67**	**86**	**93**	**96**	**11**	**4**
GSM EFR						
EXC	42	68	83	90	17	10
SP	46	77	91	94	14	6
MOD	34	48	70	90	20	10
COMB-ESM	48	73	87	92	16	8
Proposed	**67**	**87**	**93**	**96**	**11**	**4**
CELP						
EXC	28	55	76	86	24	14
SP	37	68	85	93	19	7
MOD	27	45	67	84	22	16
COMB-ESM	29	62	82	91	22	9
Proposed	**41**	**72**	**88**	**92**	**16**	**8**
MELP						
EXC	29	53	73	85	22	15
SP	45	71	87	92	16	8
MOD	33	48	71	87	21	13
COMB-ESM	37	63	79	87	20	13
Proposed	**54**	**75**	**84**	**92**	**12**	**8**

detection of VOPs for the CV units is better in case of proposed method compared to the existing methods. Average deviation using proposed method is around 5 ms and 1 ms higher compared to the clean case for CELP and MELP coders, respectively, and for the GSM coders average deviation seems to be close to clean case (see Table 3.5). In the presence of coding, proposed method and method based on spectral peaks energy performing better, compared to other methods.

3.4   VOP Detection Method for Noisy Speech

In real-time environment, noise is one of the major degradation. Hence in this work we propose a method for robust detection of the vowel onset points (VOPs) under noise. Proposed method uses spectral energy at formant frequencies of the speech segments present in glottal closure region for the detection of VOPs [94]. Here we considered spectral energy at formant frequencies instead of 500–2,500 Hz band energy, because spectral energy in 500–2,500 Hz band may not be robust under noise. In general, voiced regions contain most of the spectral energy. Within voiced region, in each pitch cycle, speech energy is dominant in glottal closure phase compared to glottal open phase. This is due to instant of significant excitation at the instant of glottal closure. Within glottal closure region, most of the energy is concentrated at the formant frequencies. Therefore in our proposed method, we have focused on spectral energy at the formant frequencies of the speech signal present in glottal closure region for the robust detection of VOPs. Formants in the glottal closure region are extracted using group delay based method [95], and the details of formant extraction method are described in the following subsection.

3.4.1   Formant Extraction Using Group Delay Function

Formant extraction from short segments of speech signal using group delay functions presented in [95, 96] is used in this work. Short segmental analysis based on conventional spectral methods suffer from the problem of poor resolution in the frequency domain. Hence high resolution property of group delay can be used for extracting formant frequencies from short segments of speech [95]. Group delay $(\tau_g(\omega))$ is defined as

$$\tau_g(\omega) = -\frac{d\phi(\omega)}{d\omega} \tag{3.13}$$

where $\phi(\omega)$ is phase function and ω is frequency variable. τ_g can be computed directly from signal $x(n)$ as

$$\tau_g(\omega) = \frac{X_i(\omega)X_r{}'(\omega) + X_r(\omega)X_i{}'(\omega)}{X_r(\omega)^2 + X_i(\omega)^2} \tag{3.14}$$

where $X_i(\omega)$ and $X_r(\omega)$ are the imaginary and real parts of Fourier transform of $x(n)$, and $X_i{}'(\omega)$ and $X_r{}'(\omega)$ are their derivatives.

$$\tau_g(\omega) \propto |X(\omega)|^2 \tag{3.15}$$

Fig. 3.9 Numerator $g(\omega)$ of
group delay function
computed for voiced speech
segment of 3 ms duration

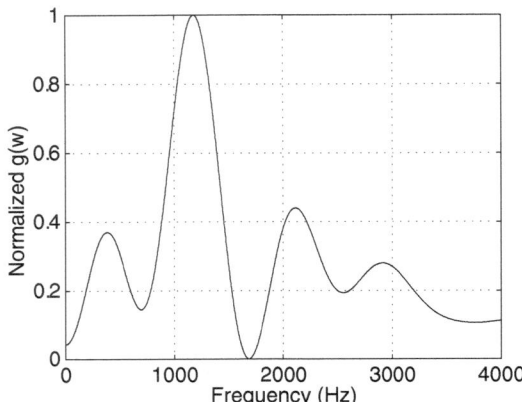

It is known that group delay function $\tau_g(\omega)$ of signal around resonant frequency is proportional to square of the magnitude of Fourier transform $|X(\omega)|^2$. Issues associated with the calculation of group delay function are primarily due to zeros present in the denominator term of Eq. (3.14). The denominator term corresponds to the magnitude spectrum of the signal, which is typically large around the formant locations, and hence it reduces the value of numerator around the formant locations. Therefore, numerator $(g(\omega))$ of group delay function is considered for the calculation of formants from short speech segments. Numerator of group delay function $g(\omega)$ is

$$g(\omega) = X_i(\omega)X_r{}'(\omega) + X_r(\omega)X_i{}'(\omega) \tag{3.16}$$

$$g(\omega) \propto |X(\omega)|^4 \tag{3.17}$$

At resonant frequencies $g(\omega)$ is proportional to $|X(\omega)|^4$, so $g(\omega)$ gives sharper peaks at resonances than $\tau_g(\omega)$. Numerator $g(\omega)$ of group delay function computed for a voiced speech segment of 3 ms duration is shown in Fig. 3.9. The peaks in the $g(\omega)$ signal correspond to the formant locations.

If we synchronize the analysis windows with the GCIs, the variation in the configuration of the vocal tract can be captured through the variation in the formant frequencies from one pitch cycle to another [96]. Formant extraction using group delay-based method is carried out with the following sequence of steps [95, 96]:

1. Consider a speech segment present in the glottal closure phase.
2. Filter the segment of the speech signal using a half Hanning window of length less than pitch period.
3. Compute the $g(\omega)$ function.
4. Pick the largest N number of peaks in the computed $g(\omega)$ function.
5. Repeat steps (1–4) at all glottal closure instants.

3.4.2 Sequence of Steps in the Proposed VOP Detection Method for Noisy Speech

The sequence of steps in the proposed method are similar to the method described in Sect. 3.2.2, and they are:

1. Determine the epoch locations (glottal closure instants) by using ZFF method.
2. Compute formants using group delay based method [95] for the speech samples present in 30% of glottal cycle starting from the GCI.
3. Determine the sum of spectral energies present at first three formant frequencies.
4. Spectral energy computed in step 3 is plotted as a function of time. Fluctuations in the spectral energy contour are smoothed by using mean smoothing with 50 ms window.
5. The change at the VOP present in the smoothed spectral energy is enhanced by computing its slope using a FOD.
6. The significant changes in the spectral characteristics present in the enhanced version of the smoothed spectral energy are detected by convolving with FOGD operator of length 100 ms.
7. Positive peaks in the proposed VOP evidence plot represent the VOP locations.

The output of each of the steps in the proposed VOP detection method is shown in Fig. 3.10. Figure 3.10a shows the speech signal /"*Don't ask me to carry an oily rag like that*"/ with manually marked VOPs. Smoothed signal of spectral energy at formant frequencies around epoch location is shown in Fig. 3.10b (step 4). Figure 3.10c shows the enhanced plot of Fig. 3.10b (step 5). VOP evidence plot obtained from the proposed method is shown in Fig. 3.10d (step 6). We can observe that manual marked VOPs in Fig. 3.10a and detected VOPs marked in Fig. 3.10d are close to each other.

3.5 Performance of the VOP Detection Method in the Presence of Background Noise

In this work, white and vehicle noise samples from Noisex-92 [97] database are considered for analysis. The noise samples are added to clean speech data to generate noisy speech data at different signal-to-noise ratio (SNR) values. Robustness of the proposed VOP detection method compared to COMB-ESM method is illustrated in Fig. 3.11 by using white noise added (SNR of 10 dB) speech utterance /"*Don't ask me to carry an oily rag like that*"/. Figure 3.11a shows the speech signal with manually marked VOPs. VOP evidence plots for the speech signal shown in Fig. 3.11a by using COMB-ESM and proposed methods are shown in Fig. 3.11b, c, respectively. From the Fig. 3.11b, c, we can observe that four spurious VOPs are detected in case of COMB-ESM VOP evidence plot, and only one spurious VOP is detected in case of proposed VOP evidence plot.

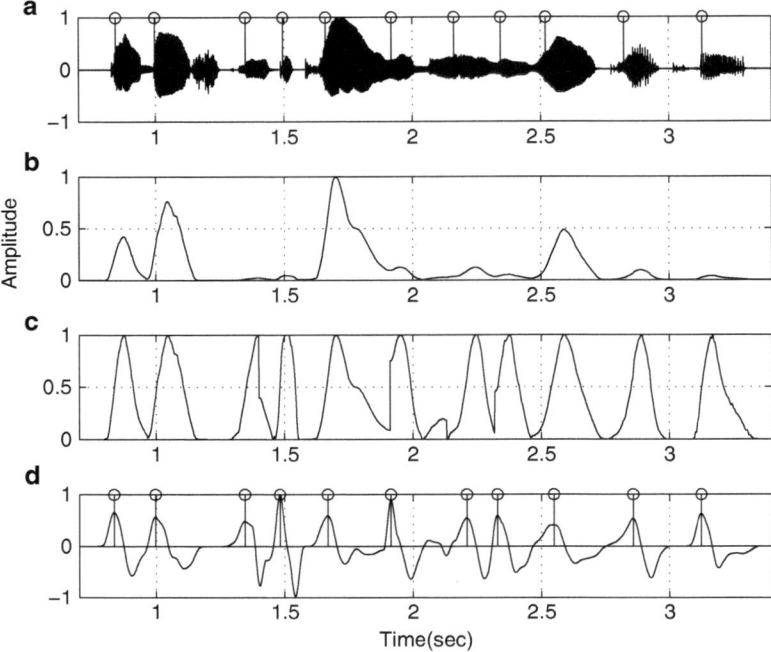

Fig. 3.10 VOP detection using proposed method for a speech utterance /"*Don't ask me to carry an oily rag like that*"/. (**a**) speech signal, (**b**) mean smoothed plot of spectral energy at formant frequencies around each epoch, (**c**) enhanced spectral energy signal, (**d**) proposed VOP evidence plot

3.5.1 VOP Detection from Continuous Speech Under Noise

Performance of the proposed VOP detection method for noisy speech is compared with COMB-ESM method and proposed VOP detection method for coded (PROP-COD) speech. Table 3.6 shows the performance of VOP detection methods on TIMIT database under noise. Column-1 in Table 3.6 indicates the VOP detection methods consider in this study. Columns 2–5 and 9–12 in Table 3.6 indicate the percentage of VOPs detected in 10, 20, 30, and 40 ms deviation. Columns 6 and 13 indicate the average deviation with respect to the manual marked VOPs. Similarly columns 7 and 8, and 14 and 15 in Table 3.6 show the percentage of miss and spurious VOPs. Results indicate that performance of the proposed method is superior compared to COMB-ESM and PROP-COD methods. From the results, it can be observed that VOP detection performance is decreasing due to noise [98]. In case of COMB-ESM and PROP-COD methods, number of spurious detections are very high due to noise at low SNR values (see Table 3.6). Poor performance of PROP-COD method under noise indicates that spectral energy at 500–2,500 Hz

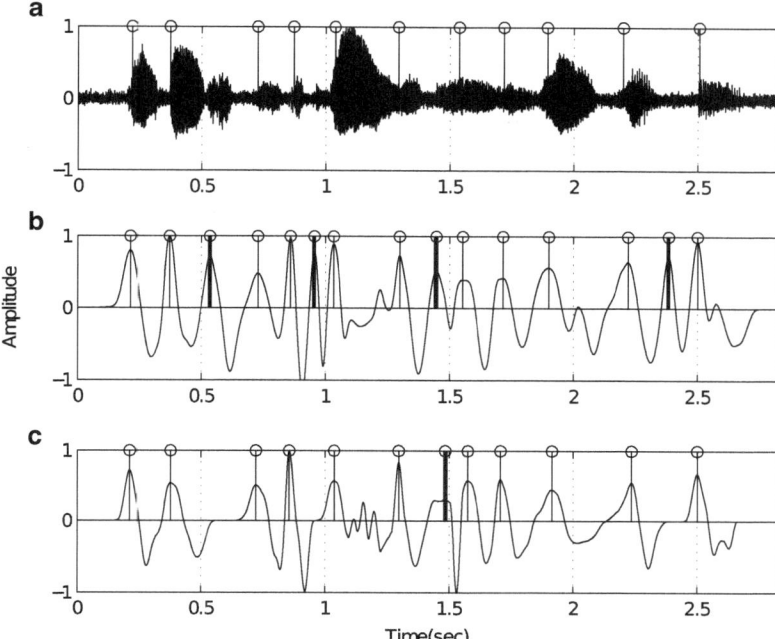

Fig. 3.11 VOP detection using COMB-ESM and proposed VOP detection methods for the white noise added (SNR of 10 dB) speech utterance /"*Don't ask me to carry an oily rag like that*"/. (**a**) speech signal with manually marked VOPs, (**b**) VOP evidence plot using COMB-ESM method, and (**c**) VOP evidence plot using proposed method

band is not robust to noise. Spurious VOPs are reduced significantly in the presence of noise by using proposed method. This is because of exploiting the high SNR characteristics present at the formant frequencies in the glottal closure phase.

3.5.2   VOP Detection from CV Units Under Noise

Table 3.7 shows the detection accuracy of VOPs using the different VOP detection methods [15] under noise by using CV units from broadcast news speech corpus (950 VOPs are considered). From the results, it is observed that proposed method is outperformed over COMB-ESM and PROP-COD methods in terms of accuracy in VOP detection (time resolution and average deviation).

Table 3.6 Performance of the VOP detection on TIMIT DATABASE (1,197 genuine VOPs) using proposed, PROP-COD, and COMB-ESM VOP detection methods under noise

White noise

VOP detection method	VOPs detected within ms (%)				AVG dev. (ms)	MISS VOPs (%)	SPU VOPs (%)
	10	20	30	40			
Clean							
COMB-ESM	51	59	70	95	16	5	3
PROP-COD	65	83	91	95	12	5	2
Proposed	**65**	**83**	**91**	**95**	**12**	**5**	**2**
0 dB							
COMB-ESM	39	49	62	85	24	15	39
PROP-COD	30	34	56	79	26	21	52
Proposed	**58**	**71**	**84**	**90**	**19**	**10**	**7**
5 dB							
COMB-ESM	42	52	65	88	22	12	32
PROP-COD	39	47	61	85	22	15	45
Proposed	**60**	**72**	**85**	**94**	**14**	**6**	**4**
10 dB							
COMB-ESM	45	58	68	92	17	8	30
PROP-COD	44	54	65	90	17	10	39
Proposed	**62**	**77**	**85**	**94**	**11**	**6**	**3**
20 dB							
COMB-ESM	48	59	71	94	16	6	18
PROP-COD	51	62	73	95	16	5	20
Proposed	**64**	**77**	**86**	**96**	**10**	**4**	**2**

Vehicle noise

VOP detection method	VOPs detected within ms (%)				AVG dev. (ms)	MISS VOPs (%)	SPU VOPs (%)
	10	20	30	40			
Clean							
COMB-ESM	51	59	70	95	16	5	3
PROP-COD	65	83	91	95	12	5	2
Proposed	**65**	**83**	**91**	**95**	**12**	**5**	**2**
0 dB							
COMB-ESM	43	54	64	89	21	11	32
PROP-COD	36	49	61	84	22	16	44
Proposed	**59**	**74**	**85**	**92**	**15**	**8**	**6**
5 dB							
COMB-ESM	46	57	68	93	19	7	28
PROP-COD	42	52	63	89	19	11	35
Proposed	**62**	**75**	**86**	**94**	**15**	**6**	**4**
10 dB							
COMB-ESM	50	59	70	94	17	6	25
PROP-COD	48	54	67	93	17	7	32
Proposed	**64**	**75**	**86**	**95**	**11**	**5**	**4**
20 dB							
COMB-ESM	52	60	71	96	16	4	18
PROP-COD	54	63	74	97	16	3	10
Proposed	**65**	**81**	**89**	**96**	**11**	**4**	**2**

Table 3.7 Performance of the VOP detection on CV units from Telugu broadcast news speech corpus (950 VOPs are considered) using proposed, PROP-COD, and COMB-ESM VOP detection methods under noise

VOP detection method	VOPs detected within ms ($\approx$ %)				AVG dev. ($\approx$ ms)	VOPs detected within ms ($\approx$ %)				AVG dev. ($\approx$ ms)
	10	20	30	40		10	20	30	40	
	Clean					Clean				
COMB-ESM	50	74	89	95	15	50	74	89	95	15
PROP-COD	69	88	93	97	11	69	88	93	97	11
Proposed	**69**	**89**	**92**	**97**	**11**	**69**	**89**	**92**	**97**	**11**
	White noise					**Vehicle noise**				
	0 dB					0 dB				
COMB-ESM	40	63	80	86	23	44	56	67	90	20
PROP-COD	38	60	78	84	23	42	55	64	88	21
Proposed	**52**	**74**	**82**	**88**	**19**	**61**	**72**	**83**	**92**	**14**
	5 dB					5 dB				
COMB-ESM	40	65	82	88	21	49	59	71	93	19
PROP-COD	39	62	81	86	21	47	56	70	92	19
Proposed	**60**	**82**	**90**	**93**	**14**	**64**	**77**	**88**	**94**	**13**
	10 dB					10 dB				
COMB-ESM	47	73	88	94	17	52	61	72	94	16
PROP-COD	45	72	86	92	17	51	60	71	92	16
Proposed	**65**	**87**	**94**	**96**	**10**	**66**	**77**	**88**	**95**	**10**
	20 dB					20 dB				
COMB-ESM	48	74	89	95	17	55	66	74	96	16
PROP-COD	45	72	86	92	17	51	60	71	92	16
Proposed	**68**	**88**	**93**	**96**	**10**	**66**	**83**	**87**	**96**	**10**

3.6 Summary

In this chapter, we have proposed methods for robust detection of VOPs in the presence of speech coding and background noise conditions. Proposed methods are based on spectral energies of the speech segments present in the glottal closure region. The reasons for choosing the speech segments at the glottal closure region for deriving the spectral energy are (1) speech signal during glottal closure phase has high SNR characteristics, and (2) vocal tract resonances during glottal closure phase are more accurate. These merits are exploited in the proposed VOP detection methods by considering 30% of glottal cycle starting from glottal closure instant instead of conventional 20 ms frame with block processing. From the results, it is observed that performance of the proposed VOP detection methods is superior compared to existing methods under clean, coded, and noisy conditions.

VOP detection performance under coding is studied for GSM FR, GSM EFR, CELP, and MELP coding environments. From the results, it is observed that reduction in VOP detection performance is not significant in case of GSM coders and significant in case of CELP and MELP coders. It is observed that VOP detection performance from CELP-coded speech is decreased compared to MELP coder, even though bit rate of MELP coder is less than CELP coder. As speech coders preserve the characteristics of vocal tract system, VOP detection methods based on spectral characteristics performed better compared to other methods. In case of both clean and coded speech, proposed VOP detection method is giving improved performance compared to excitation source, spectral peaks energy, modulation spectrum, and COMB-ESM methods. Performance of the VOP detection under noise is severely affected due to spurious VOPs at low SNR values. Spurious VOPs are reduced significantly by using proposed method, which uses the spectral energy at first three formants. It is also observed that performance of the proposed VOP detection methods for coding and noise is nearly same for clean case.

Chapter 4
Consonant–Vowel Recognition in the Presence of Coding and Background Noise

Abstract In this chapter, an approach for improving the recognition performance of CV units under clean, coded, and noisy conditions is presented. Proposed CV recognition method is carried out in two stages. In the first stage vowel category of CV unit is recognized, and in the second stage consonant category is recognized. At each stage of the proposed method, complementary evidences from support vector machine (SVM) and hidden Markov models (HMM) are combined for enhancing the recognition performance of CV units. In the proposed CV recognition approach, VOP is used as an anchor point for extracting features from the CV unit. Therefore, VOP detection methods presented in previous chapter are used for this work. Performance of the proposed CV recognition method is demonstrated under coding and noisy conditions. Recognition studies are carried out using isolated CV and CV units from Telugu broadcast news databases. Further, performance of the CV recognition system under background noise is improved by using combined temporal and spectral processing-based preprocessing methods.

In Chap. 3, robust VOP detection methods are discussed, and their performance is evaluated using isolated CV utterances and CV utterances continuous speech. This chapter discusses about CV recognition applications based on vowel onset points. Here, CV recognition performance is analyzed in the presence of coding and background noisy environments. In each application, the impact of accuracy of VOP detection on CV recognition performance is analyzed. This chapter is organized as follows. In Sect. 4.1, speech databases and CV units considered in this work are described. Section 4.2 presents the proposed CV recognition method. Impact of the accuracy in VOP detection on CV recognition performance is discussed in Sect. 4.3. In Sect. 4.4, performance of the proposed CV recognition approach is analyzed under coding and compared with the existing methods. Performance of the CV recognition system under background noise conditions is presented in Sect. 4.5. In Sect. 4.6, combined temporal and spectral preprocessing technique

K.S. Rao and A.K. Vuppala, *Speech Processing in Mobile Environments*, SpringerBriefs in Electrical and Computer Engineering, DOI 10.1007/978-3-319-03116-3_4, © Springer International Publishing Switzerland 2014

and its effectiveness for CV recognition under background noise conditions are presented. Summary of the issues discussed in this chapter is given in Sect. 4.7.

4.1 Consonant–Vowel Unit Databases

In this work, two CV unit databases are used for analyzing the performance of CV recognition system. Database-1 consists of isolated CV utterances in Indian languages collected by International Institute of Information Technology, Hyderabad, India [99]. Among different CV units present in the Database-1, 145 most frequently occurred CV units are considered for this study and they are shown in Table 4.1. It is

Table 4.1 List of 145 CV units

Manner of articulation	Place of articulation	Vowel /a/	/i/	/u/	/e/	/o/
Unvoiced	Velar	ka	ki	ku	ke	ko
Unaspirated	Palatal	cha	chi	chu	che	cho
(UVUA)	Alveolar	Ta	Ti	Tu	Te	To
	Dental	ta	ti	tu	te	to
	Bilabial	pa	pi	pu	pe	po
Unvoiced	Velar	kha	khi	khu	khe	kho
aspirated	Palatal	Cha	Chi	Chu	Che	Cho
(UVA)	Alveolar	Tha	Thi	Thu	The	Tho
	Dental	tha	thi	thu	the	tho
	Bilabial	pha	phi	phu	phe	pho
voiced	Velar	ga	gi	gu	ge	go
Unaspirated	Palatal	ja	ji	ju	je	jo
(VUA)	Alveolar	Da	Di	Du	De	Do
	Dental	da	di	du	de	do
	Bilabial	ba	bi	bu	be	bo
voiced	Velar	gha	ghi	ghu	ghe	gho
aspirated	Palatal	jha	jhi	jhu	jhe	jho
(VA)	Alveolar	Dha	Dhi	Dhu	Dhe	Dho
	Dental	dha	dhi	dhu	dhe	dho
	Bilabial	bha	bhi	bhu	bhe	bho
Nasals	Dental	na	ni	nu	ne	no
	Bilabial	ma	mi	mu	me	mo
Semivowels	Palatal	ya	yi	yu	ye	yo
	Alveolar	ra	ri	ru	re	ro
	Dental	la	li	lu	le	lo
	Bilabial	va	vi	vu	ve	vo
Fricatives	Velar	ha	hi	hu	he	ho
	Alveolar	sha	shi	shu	she	sho
	Dental	sa	si	su	se	so

difficult to recognize short and long vowels in the continuous speech. Hence, both short and long vowels are considered as one vowel only. Simple language model will take care of short and long vowels during speech recognition. Database-1 contains utterances of 25 males and 25 females from different parts of India, in four different sessions. For each CV unit around 400 utterances are available in the database. In that 322 utterances are used for training and 78 utterances are used for testing.

Database-2 is Telugu broadcast news corpus described in Sect. 3.1.2. Among 20 news bulletins available in the database, 15 bulletins (8 males + 7 females) are used for training and 5 bulletins (3 males + 2 females) are used for testing. Manual marked syllable boundaries are available in Database-2. In this work, among 145 prominent CV classes, 95 CV classes (see bold ones in Table 4.1) whose frequency of occurrence more than 50 are considered for the analysis. Total number of CV utterances considered in this study are 52,703 (38,729 are used for training and 13,974 are used for testing), and their contribution is more than 95% of CV units present in the Database-2.

4.2   Two-Stage CV Recognition System

High similarity and large number of CV classes are the major issues involved in CV recognition. In this work, we proposed two-stage CV recognition method for the recognition of CV units in Indian languages. In the first stage vowel will be recognized, and in the second stage consonant will be recognized. In both stages, evidences from HMM and SVM models are combined with appropriate weights [100]. In Sect. 4.2.1, the motivations for the proposed method are described. Proposed approach for CV recognition is presented in Sect. 4.2.2. The framework used to carry out CV recognition studies is described in Sect. 4.2.3. In Sect. 4.2.4, we analyzed the performance of proposed CV recognition approach.

4.2.1   Motivations for the Proposed CV Recognition Approach

- Single-stage models may not be appropriate for classification of large number and highly confusable CV classes. Therefore, two-stage approach is proposed. In the proposed approach, 145 classes are divided into 5 subgroups based on vowel category (columns 3–7 of Table 4.1).
- For enhancing the recognition performance, hybrid models are explored at each stage to capture the CV characteristics in different ways. In this work the hybrid model consists of combination of HMM and SVM. It is known that HMMs capture distribution and sequential knowledge from the feature vectors of the specific class. SVMs are known for capturing discriminative characteristics between the desired class and rest of the classes by using positive and negative examples from the desired class and rest of the classes, respectively. Since HMM

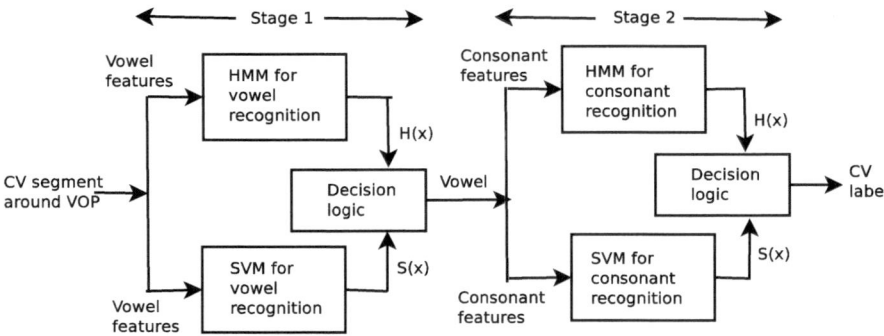

Fig. 4.1 Two-stage CV recognition system using HMM and SVM

and SVM models capture the class-specific knowledge based on different modalities, combining the evidence from these models may improve the recognition performance.

• In the existed multilayer CV recognition approach [54], same features extracted around VOP are used at each stage for developing the acoustic models. But, in the proposed approach different features are used at each stage by using VOP as an anchor point. Recognition performance of the proposed method depends on the accuracy of the feature extraction associated with vowel, consonant, and transition regions of CV units. Therefore, we used proposed accurate VOP detection methods described in Chap. 3 for determining different regions of CV units.

4.2.2 Proposed CV Recognition Approach

Proposed two-stage CV recognition model is shown in Fig. 4.1. In the proposed method, at each stage decision is taken by combining evidences from SVM and HMM using below equation.

$$C(X) = arg\ max_i\ (w1 * S_i(X) + w2 * H_i(X)) \tag{4.1}$$

where $S_i(X)$ and $H_i(X)$ correspond to normalized evidence scores from the SVM and HMM models, respectively, for test utterance X and i is the class identity. C is hypothesized class. $w1$ and $w2$ ($w2 = 1 - w1$) are the weights given for SVM and HMM evidence scores, respectively. In our study, $w1$ is varied in steps of 0.02 from 0 to 1.

4.2.3 Framework

Mel-Frequency Cepstral Coefficients (MFCC) [101] extracted from every 20 ms of CV segment with 5 ms frame shift are used for developing the acoustic models. The steps involved in extraction of MFCC features from speech signals are described in Appendix A. In the proposed CV recognition approach different features are used at each stage by using VOP as an anchor point. At the first stage, vowel recognition models are developed by extracting features from VOP to end of the CV segment (i.e., only vowel region), and at the second stage consonant models are developed using features extracted from beginning of the CV segment to the transition region. In this study 40 ms speech segment following VOP is considered as transition region [99]. VOP detection methods proposed in previous chapter are used to carry out this study.

HMM is a stochastic signal model with finite set of states, each of which is associated with a probability distribution. Transitions among the states are governed by a set of probabilities known as transition probabilities. In a particular state an outcome or observation can be generated, according to the associated probability distribution (more details are given in Sect. C.1). In this study, HMM models are trained by using maximum likelihood approach. HMM Tool Kit (HTK) [102] is used for developing the HMM models. MFCC feature vectors of size 39 dimension (13 MFCC + delta + delta–delta coefficients) are used for developing HMM models. In the proposed CV recognition approach, vowel and consonant HMM models are developed by using 3 states and 64 mixers per state. Performance of proposed method is compared with single-stage HMM models which are built using 4 states and 64 mixtures.

SVMs are designed for binary classification. Multi-class (n-class) classification issues can be solved by using combination of binary support vector machines. One-against-the-rest approach is used for decomposing the n-class classification problem into binary classification problems (more details are given in Sect. C.2). Open source SVMTorch [103] is used for developing the SVM models. Fixed pattern length (PL) of 10 and Gaussian kernel of width 40 are used to build SVM models for proposed CV recognition approach and single-stage SVM approach. Fixed pattern length PL is obtained from variable segment length (SL) utterances. If segment length SL is greater than PL, few frames of the segment are omitted. If the segment length SL is smaller than PL, few frames of the segment are repeated. The dimension of feature vector extracted from each utterance for developing the SVM models is 390 (10×39 MFCCs per frame).

4.2.4 Performance of the CV Recognition System

The effectiveness of proposed CV recognition approach is evaluated by performing experiments using isolated CV units from Database-1. In Fig. 4.2 different evidences for test utterance from class 26 ($/t/$ class) are shown for the recognition of

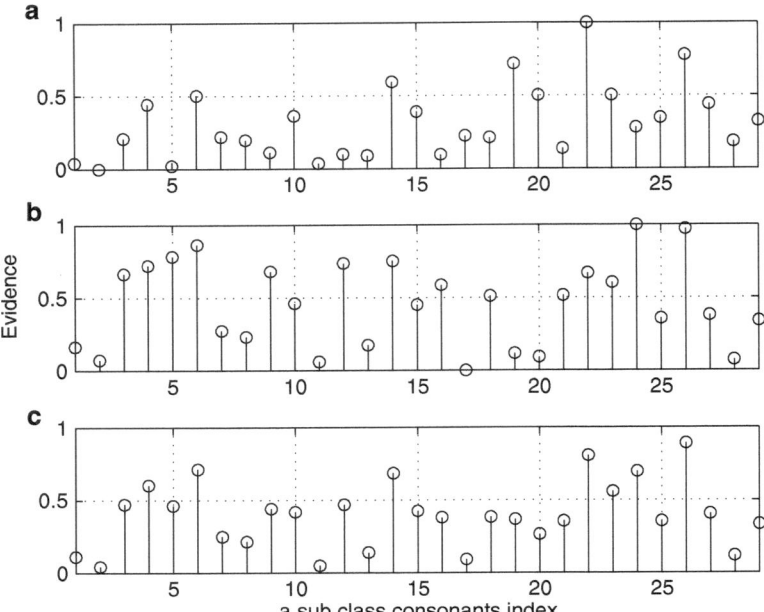

Fig. 4.2 Evidence plots for test utterance /t/ (**a**) SVM evidence. (**b**) HMM evidence. (**c**) HMM+SVM evidence

consonant from /a/ subclass. Evidences for SVM, HMM, and SVM+HMM models for the utterance /t/ are shown in Fig. 4.2a–c, respectively. From Fig. 4.2, it is observed that highest evidences have occurred for the classes 22 and 24 from SVM and HMM models, respectively, for class 26 test utterance. It is seen that class 26 has the largest evidence from combination of HMM and SVM evidences. Hence this behavior of combination of complimentary evidences from HMM and SVM may improve the recognition performance of CV units.

In the proposed approach, the evidences of SVM and HMM are combined using the weighting rule. The weighting factor of SVM evidence $w1$ is varied from 0 to 1, in steps of 0.02. With this we get a total of 51 combinations of weighting factors. The vowel recognition performance of the combined system for various combinations of the weighting factors is shown in Fig. 4.3. It is observed that the best recognition performance is about 97 % for the weighting factors 0.64 and 0.36 for the confidence scores of SVM and HMM, respectively (see Fig. 4.3).

Figure 4.4 shows the recognition performance of consonants of /a/ vowel subgroup using proposed method by combining SVM and HMM evidences with different weights between 0 and 1, in steps of 0.02. Highest recognition performance of about 74 % is observed for the weighting factors 0.38 and 0.62 for the confidence scores of SVM and HMM, respectively (see Fig. 4.4).

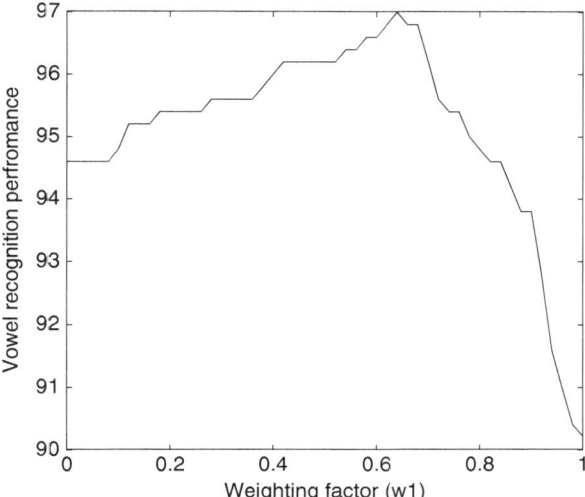

Fig. 4.3 Recognition performance of vowel with different values of weight w1

Fig. 4.4 Recognition
performance of /a/ vowel
subgroup consonant with
different values of weight w1

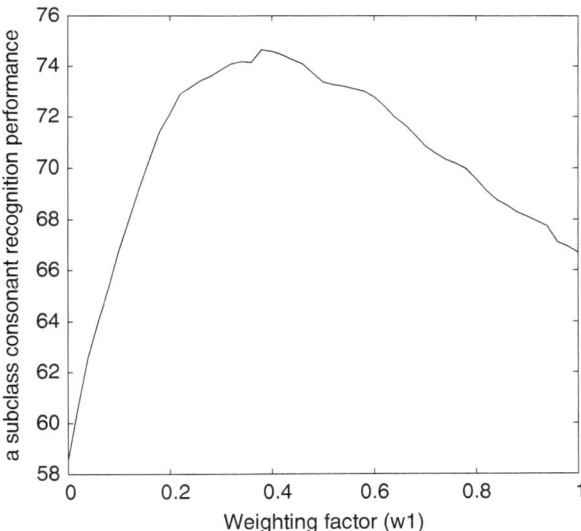

Table 4.2 shows the vowel and consonant recognition performance of HMM, SVM, and SVM+HMM models using isolated CV units. From the results, it is observed that the performance of HMM models seems to be slightly superior compared to SVM models for vowel recognition. This is due to the fact that, HMMs are good at capturing the state sequence corresponding to the sequence of vocal tract shapes. The sequences of vocal tract shapes are distinct for each vowel. From the results, it is also observed that the performance of SVM models

Table 4.2 Recognition performance of vowel and consonant from isolated CV units using SVM, HMM, and SVM+HMM acoustic models

Recognition system	Vowel recognition performance (%)
SVM	90.22
HMM	94.6
SVM+HMM	**97**
	Consonant recognition performance (%)
SVM	58.72
HMM	50.82
SVM+HMM	**70.9**

Table 4.3 Recognition performance of 145 CV units using single-stage SVM, HMM, SVM+HMM, two-stage SVM, HMM, and proposed approaches

Recognition system	Recognition performance ($\approx$ %)
Single-stage SVM	46.19
Single-stage HMM	37.42
Single-stage SVM+HMM	53.22
Two-stage SVM	52.98
Two-stage HMM	48.07
Proposed (Two-stage SVM+HMM)	**68.77**

seems to be superior compared to HMM models for consonant recognition. This is because SVM models are trained using one-against-rest-approach to capture the discriminative information present in highly similar and confusable consonant classes. There is an improvement of around 4% for vowel recognition and 11% for consonant recognition by using proposed approach.

Performance of the proposed two-stage CV recognition method is compared with different single- and two-stage HMM and SVM models, and results are shown in Table 4.3. From the results, it is observed that the performance of CV recognition using proposed two-stage approach is superior compared to other approaches. From this study, it is observed that performance of two-stage models is superior compared to single-stage models. Combination of SVM and HMM evidences in both single-stage and proposed CV recognition methods has shown significant improvement in the recognition performance (see rows 4 and 7 in Table 4.3).

4.3   Impact of Accuracy in VOP Detection on CV Recognition

Impact of accuracy in VOP detection on proposed CV recognition system is analyzed by performing studies on recognition of CV units from Telugu broadcast database in the presence of coding. Table 4.4 shows the recognition performance

Table 4.4 Recognition performance of CV units by the proposed CV recognition method using COMB-ESM and proposed VOP detection methods

VOP method	Vowel recognition performance (%)				
	Clean	GSM FR	GSM EFR	CELP	MELP
COMB-ESM	91.48	90.21	90.61	88.42	89.91
Proposed VOP	**91.92**	**90.42**	**91.02**	**89.54**	**90.22**
	Consonant recognition performance (%)				
COMB-ESM	70.41	68.97	69.22	61.82	60.47
Proposed VOP	**72.58**	**71.44**	**71.66**	**66.27**	**64.19**
	CV recognition performance (%)				
COMB-ESM	64.41	62.22	62.72	54.66	54.37
Proposed VOP	**66.72**	**64.6**	**65.22**	**59.34**	**57.92**

of vowels, consonants, and CV units for the proposed CV recognition method by using COMB-ESM (described in Sect. 2.1) and proposed (described in Sect. 3.2) VOP detection methods [15, 87, 94, 104]. In Table 4.4, column-1 indicates the VOP detection method. Columns 2–6 indicate recognition performance for clean, GSM FR, GSM EFR, CELP, and MELP coders under matched condition (trained and tested under the similar condition). From the results, it is observed that recognition performance of CV units is increased (up to 5%) by using proposed VOP detection method compared to COMB-ESM method. Improvement in performance is more significant (around 5%) in case of CELP coding. Impact of accuracy of VOP detection is significant in case of consonant recognition compared to vowel recognition.

4.4 Performance of CV Recognition System Under Coding

4.4.1 Isolated CV Units Recognition Under Coding

Table 4.5 shows the vowel and consonant recognition performance of HMM, SVM, and SVM+HMM models under clean and coded conditions using isolated CV units. In Table 4.5, column-1 indicates the acoustic model used for developing the system. Columns 2–6 indicate the recognition performance for clean GSM FR, GSM EFR, CELP, and MELP cases, respectively, using acoustic models trained with clean speech. Columns 7–10 show the recognition performance under matched condition (trained and tested with corresponding speech) for GSM FR, GSM EFR CELP and MELP coded cases, respectively. From the results, it is observed that the combination of evidences has shown improvement over individual evidences. Recognition performance has improved by 5 % and 14 % for vowel and consonant recognition, respectively. From Table 4.5, we can observe that the coding effect on vowel and

Table 4.5 Recognition performance of vowel and consonant from isolated CV units using SVM, HMM, and SVM+HMM acoustic models under different coders

Recognition system	Vowel recognition performance (%)								
	Clean training					Matched training			
	Clean	GSM FR	GSM EFR	CELP	MELP	GSM FR	GSM EFR	CELP	MELP
SVM	90.22	84	87.2	83.4	82.6	89.2	89.6	86.8	87.6
HMM	94.6	92	92.4	87.2	91.6	93.2	93.6	91	92.2
SVM+HMM	**97**	**92.8**	**94.2**	**91.6**	**93**	**96.6**	**96.6**	**95**	**97**
	Consonant recognition performance (%)								
SVM	58.72	52.75	55.17	41.75	39.81	56.85	57.14	53.69	50.34
HMM	50.82	46.85	47.32	36.81	30.8	48.75	48.77	43.35	40.34
SVM+HMM	**70.9**	**66.75**	**67.14**	**53.6**	**51.7**	**69.12**	**69.16**	**64.47**	**60.2**

Table 4.6 Recognition performance of 145 CV units using single- and two-stage approaches under different coders

Recognition system	Recognition performance (%)								
	PCM (clean) training					Matched training			
	PCM	GSM FR	GSM EFR	CELP	MELP	GSM FR	GSM EFR	CELP	MELP
Single-stage SVM	46.19	34.92	37.44	26.54	29.06	39.12	41.02	36.21	38.54
Single-stage HMM	37.42	28.34	29.85	21.06	22.61	33.86	34.65	30.12	31.17
Single-stage SVM +HMM	53.22	42.36	43.91	33.42	36.13	50.03	50.86	44.16	46.24
Two-stage SVM	52.98	44.31	48.11	34.82	32.88	50.71	51.19	46.60	44.09
Two-stage HMM	48.07	43.10	43.72	32.09	28.21	45.43	45.65	39.45	37.19
Proposed	**68.77**	**61.94**	**63.25**	**49.09**	**48.08**	**66.77**	**66.81**	**61.25**	**58.4**

consonant recognition is minimized by using hybrid (SVM+HMM) models under matched condition. The effect of GSM coders on recognition performance is almost nullified using the proposed method under matched condition. In the case of CELP and MELP coders also, the performance is significantly improved by using proposed method under matched condition.

Performance of the proposed two-stage CV recognition method is compared with different single- and two-stage HMM and SVM models, and results are shown in Table 4.6. From the results, it is observed that the performance of CV recognition using proposed two-stage approach is superior compared to other approaches. From this study, it is observed that performance of two-stage models is superior compared to single-stage models. Combination of SVM and HMM evidences in both single-stage and proposed CV recognition methods has shown significant improvement

Table 4.7 Recognition performance of vowel and consonants of CV units from Telugu broadcast news database using SVM, HMM, and SVM+HMM acoustic models under different coders

Recognition system	Vowel recognition performance (%)								
	PCM (clean) training					Matched training			
	PCM	GSM FR	GSM EFR	CELP	MELP	GSM FR	GSM EFR	CELP	MELP
SVM	86.22	82.66	83.23	73.42	69.11	83.61	84.14	84.82	85.41
HMM	87.41	82.82	84.08	76.66	73.08	86.42	86.91	85.12	86.12
SVM+HMM	**91.92**	**86.69**	**88.41**	**80.46**	**78.14**	**90.42**	**91.02**	**89.54**	**90.22**
	Consonant recognition performance (%)								
SVM	66.42	55.17	59.32	35.75	33.23	65.06	65.22	56.46	55.23
HMM	58.34	48.46	51.92	33.39	31.20	57.96	58.01	51.17	50.75
SVM+HMM	**72.58**	**62.41**	**65.54**	**41.43**	**39.43**	**71.44**	**71.66**	**66.27**	**64.19**

in the recognition performance (see rows 6 and 9 in Table 4.6). Results are also indicating that the effect of coding on CV recognition performance is significant in case of models trained with clean speech. The effect of coding is reduced drastically by using proposed CV recognition approach under matched condition.

4.4.2 CV Units Recognition from Continuous Speech in the Presence of Coding

Performance of proposed CV recognition approach is evaluated by using CV units considered from Telugu broadcast database. In this work, 95 CV classes are considered (see bold font ones in Table 4.1). Table 4.7 shows the vowel and consonant recognition performance using HMM, SVM, and SVM+HMM models under clean and coded conditions. From the results (see Table 4.7), it is observed that the combination of evidences has shown the improvement over individual evidences. Performance improvement is around 4–5 % for vowel recognition and 6–10 % for consonant recognition. From Table 4.7, we can observe that the coding effect on vowel and consonant recognition is minimized by using proposed approach under matched condition.

Recognition performance of CV units using different approaches under coding is shown in Table 4.8. From the results, it is observed that the performance of CV recognition using proposed approach is superior compared to other approaches. Effect of coding on recognition of CV units from continuous speech is observed to be high compared to isolated CV units. Effect of coding is reduced by proposed approach under matched condition.

Table 4.8 Recognition performance of 95 CV units from Telugu broadcast news database using single- and two-stage approaches under different coders

Recognition system	Recognition performance (%)								
	PCM (clean) training					Matched training			
	PCM	GSM FR	GSM EFR	CELP	MELP	GSM FR	GSM EFR	CELP	MELP
Single-stage SVM	53.7	39.45	43.66	16.27	19.27	50.35	51.16	34.81	42.55
Single-stage HMM	44.27	33.45	38.02	19.45	23.55	42.34	42.87	35.96	40.55
Single-stage SVM +HMM	59.26	47.42	51.24	23.85	27.26	57.62	58.11	45.22	48.22
Two-stage SVM	57.27	45.60	49.37	26.25	22.97	54.39	54.88	47.89	47.17
Two-stage HMM	50.99	40.13	43.65	25.59	22.80	50.08	50.42	43.55	43.71
Proposed	**66.72**	**54.11**	**57.94**	**33.34**	**30.81**	**64.6**	**65.22**	**59.34**	**57.92**

4.5 Performance of CV Recognition System in the Presence of Background Noise

Effectiveness of proposed two-stage CV recognition approach is evaluated by using Telugu broadcast news database at different signal-to-noise ratios (SNRs) for two different noise types. In this work, white and vehicle noises from NOISEX-92 database are considered. Table 4.9 shows the vowel and consonant recognition performance under noisy environments. For this study classification models are trained with clean speech and tested with noisy speech. From the results, it is observed that the recognition performance has reduced significantly due to noise, and degradation is prominent at low SNR values. Among the two noises considered, effect of white noise is higher compared to vehicle noise. It is due to the fact that white noise affects entire speech spectrum, whereas vehicle noise effects only at low frequencies. From the results, it is observed that vowel and consonant recognition performance under noise has improved by combining the evidences from HMM and SVM models. The observed improvement in case of vowel recognition is 3–5 %, and 2–10 % in case of consonant recognition.

Performance of the proposed CV recognition system in the presence of noise is compared using existing single-stage and two-stage methods and results are shown in Table 4.10. From the results, it is observed that proposed two-stage CV recognition approach has outperformed other approaches, except at low SNR values of white noise (see rows 5–8 in Table 4.10). In the presence of noise also, combination of complimentary evidences from HMM and SVM has shown improvement in the CV recognition performance. Recognition performance of CV units is improved up to 13 % by using proposed method compared to other approaches.

Table 4.9 Recognition performance of vowel and consonants from CV units using SVM, HMM, and SVM+HMM acoustic models under different background noise cases

Recognition system	White noise (SNR levels in dB)					
	Clean	0 dB	5 dB	10 dB	20 dB	30 dB
Vowel recognition performance (%)						
SVM	86.22	20.32	28.81	36.72	72.86	80.11
HMM	87.41	21.41	33.82	44.22	77.12	81.72
SVM+HMM	**91.92**	**24.42**	**37.66**	**48.11**	**80.31**	**86.52**
	Vehicle noise					
SVM	86.22	69.34	73.66	80.88	82.66	84.96
HMM	87.41	72.82	77.66	81.72	83.12	85.82
SVM+HMM	**91.92**	**77.46**	**81.88**	**85.72**	**85.66**	**89.12**
Consonant recognition performance (%)						
	White noise					
SVM	66.42	16.20	21.12	27.66	39.41	45.20
HMM	58.34	9.81	16.46	24.06	37.17	42.67
SVM+HMM	**72.58**	**18.91**	**23.69**	**32.09**	**46.85**	**51.57**
	Vehicle noise					
SVM	66.42	28.34	37.06	45.20	54.75	58.52
HMM	58.34	27.96	35.17	41.88	50.39	53.72
SVM+HMM	**72.58**	**33.65**	**42.55**	**50.85**	**63.58**	**68.23**

Table 4.10 Overall CV recognition using different acoustic models under different background noise cases

Recognition system	Recognition performance (%)					
	White noise (SNR levels in dB)					
	Clean	0 dB	5 dB	10 dB	20 dB	30 dB
Single-stage SVM	53.70	5.85	10.12	15	28.36	41.97
Single-stage HMM	44.27	7.11	14.42	16.52	24.66	33.17
Single-stage SVM+HMM	59.26	10.24	18.07	22.12	32.52	43.08
Two-stage SVM	57.27	3.29	6.09	10.16	28.71	36.21
Two-stage HMM	50.99	2.10	5.56	10.63	28.66	34.86
Proposed	**66.72**	**4.62**	**8.92**	**15.44**	**37.63**	**44.62**
	Vehicle noise					
Single-stage SVM	53.70	15.13	23.52	32.20	43.72	47.50
Single-stage HMM	44.27	13.18	22.52	27.44	34.63	38.06
Single-stage SVM+HMM	59.26	21.36	26.36	35.17	47.62	51.12
Two-stage SVM	57.27	19.65	27.29	36.56	45.26	49.72
Two-stage HMM	50.99	20.36	27.31	34.22	41.88	46.10
Proposed	**66.72**	**26.07**	**34.84**	**43.59**	**54.46**	**60.81**

4.6   Application of Combined Temporal and Spectral Processing Methods for CV Units Recognition Under Background Noise

In this section, we demonstrate the effectiveness of speech enhancement technique proposed in [105] for speech recognition task in noisy environment. Section 4.6.1 describes the combined temporal and spectral processing (TSP) technique for enhancement of noisy speech. Section 4.6.2 presents the recognition performance of CV units under noise by using different preprocessing techniques.

4.6.1   Combined TSP Method for Enhancement of Noisy Speech

Methods developed in literature for the enhancement of noisy speech are grouped into spectral processing and temporal processing. Spectral processing methods are based on the fact that spectral values of noisy speech will have both speech and noise components [106–108]. The spectral characteristics of noise are therefore estimated and removed to obtain the enhancement. Spectral processing methods are popular due to simplicity and effectiveness. The demerit of spectral processing methods is the need for explicit modeling of spectral characteristics of noise. This is difficult for highly non-stationary noise cases. Temporal processing methods are based on identifying and enhancing the speech-specific regions of noisy speech [109–111]. The merit of temporal processing is in the enhancement of speech-specific regions and does not require any explicit modeling of degradation. The demerit may be the ineffectiveness in minimizing the degrading component, since it is not explicitly modeled. It may be possible that one domain of processing may aid other domain of processing in minimizing the demerit. Therefore, one can effectively combine temporal and spectral processing approaches to achieve improved performance [105].

In the temporal and spectral processing methods the enhancement is achieved by identifying and enhancing speech-specific features from the noisy speech present both in the temporal and in the spectral domains. The temporal processing involves identifying and enhancing the speech-specific features present at the gross and fine temporal levels. The main objective of the gross level processing is to identify and enhance the speech components at the sound units (100–300 ms) level and the objective of the fine level processing is to identify and enhance the speech-specific features at the segmental (10–30 ms) level. The high SNR speech regions at gross level are determined using speech-specific parameters like sum of 10 largest peaks in the discrete Fourier transform (DFT) spectrum, smoothed Hilbert envelope of the LP residual and modulation spectrum values from the noisy speech signal. The motivation behind using these three parameters is that they represent different aspects of the speech production mechanism. The sum of the peaks in the DFT

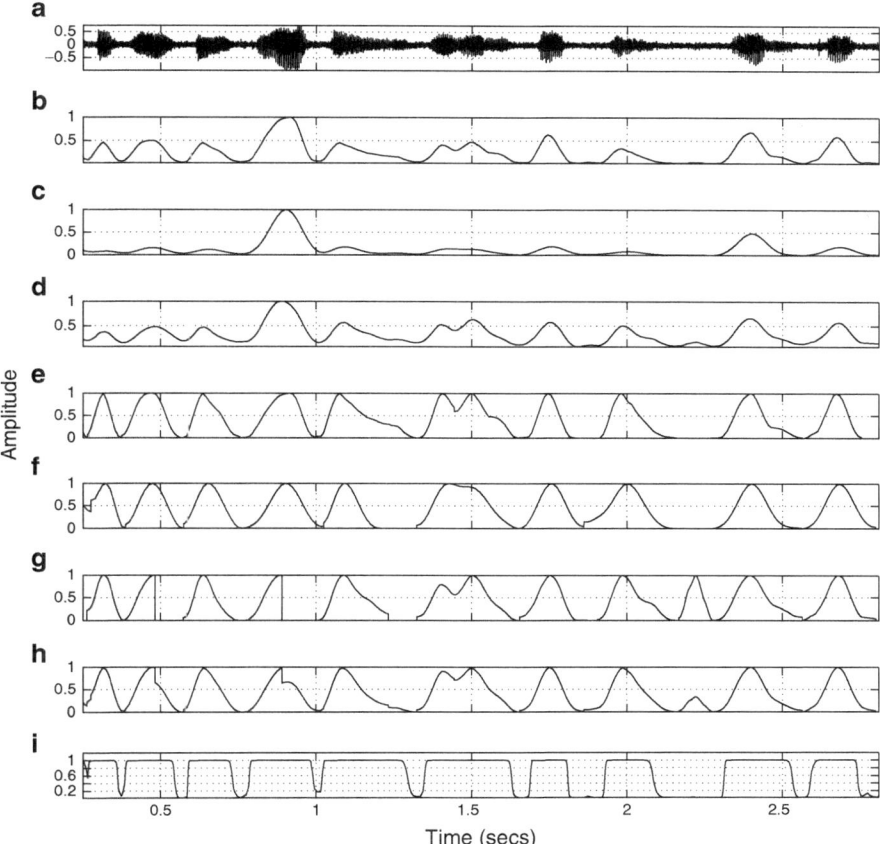

Fig. 4.5 Computation of gross weight function: (**a**) noisy speech, (**b**) sum of peaks in the DFT spectrum, (**c**) smoothed Hilbert envelope of LP residual, (**d**) modulation spectrum, (**e**) enhanced DFT spectrum values, (**f**) enhanced smoothed Hilbert envelope values, (**g**) enhanced modulation spectrum values, (**h**) normalized sum of (**e**), (**f**), and (**g**); and (**i**) gross weight function

spectrum represents predominantly the vocal tract information of speech production. The smoothed Hilbert envelope of the LP residual represents predominantly the excitation source information of speech production. The modulation spectrum represents the long-term (supra segmental) information of speech production. Since the origin of these three parameters is different, combining them may improve the robustness and also the detection accuracy as compared to any one of them.

Figure 4.5 shows the computation process of gross weight function. Figure 4.5a indicates the noisy speech signal. Figure 4.5b–d indicate the evidences derived from spectral peaks of the DFT spectrum, Hilbert envelope of the LP residual, and modulation spectrum. Figure 4.5e–g indicate the enhanced evidences of Fig. 4.5b–d. Combination of enhanced evidence of Fig. 4.5e–g is shown in Fig. 4.5h. The final

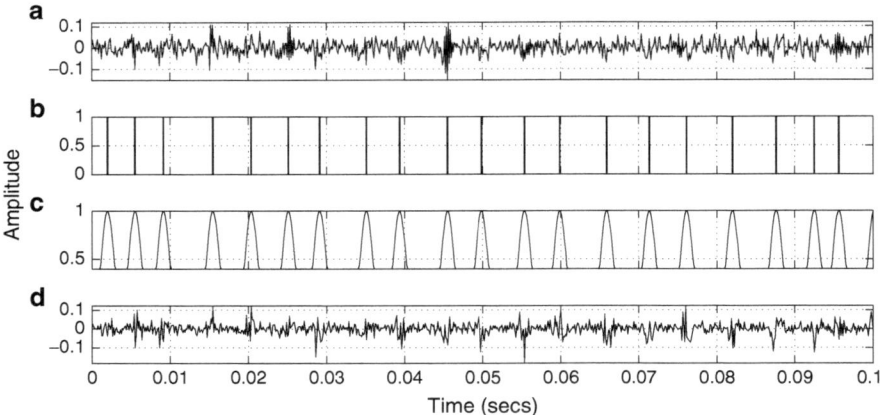

Fig. 4.6 Computation of fine weight function: (**a**) degraded LP residual, (**b**) instants of significant excitation, (**c**) fine weight function, and (**d**) enhanced LP residual

gross weight function shown in Fig. 4.5i is derived from Fig. 4.5h by applying the sigmoid nonlinear function. The gross weight function indicates the higher value during speech regions and low values during nonspeech regions. By modifying the noisy speech LP residual with the gross weight function will enhance the residual by deemphasizing the nonspeech or noisy regions.

The high SNR speech-specific features at the fine level are identified by using the knowledge of the instants of significant excitation. In this work, instants of significant excitation (epochs) are determined using zero frequency filter (ZFF)-based method [92] (see Sect. 3.2.1). The basis for the fine level temporal enhancement is that the voiced speech is produced as a result of excitation of quasi-periodic glottal pulses and unvoiced speech is produced as a result of excitation of onset of events like burst and frication. The significant excitation in each glottal cycle takes place at the instant of glottal closure. By locating the instants of significant excitation, it is possible to enhance speech around the instants relative to other regions. A weight function is derived for the LP residual from the instants of significant excitation to enhance the excitation source information around these instants relative to other regions. A final weight function is derived by combining gross and fine weight functions, which is then multiplied with the LP residual of the noisy speech signal to enhance the speech-specific features in the temporal domain.

Computation of fine weight function is shown in Fig. 4.6. Figure 4.6a shows the LP residual for a noisy speech segment. Figure 4.6b, c indicates the instants of significant excitation and fine weight function, respectively, for the noisy speech segment shown in Fig. 4.6a. Figure 4.6d shows the enhanced LP residual after applying the fine weight function over noisy LP residual shown in Fig. 4.6a. The overall temporal processing method for enhancing the speech is shown in Fig. 4.7. Figure 4.7a shows the segment of noisy LP residual. The gross weight function

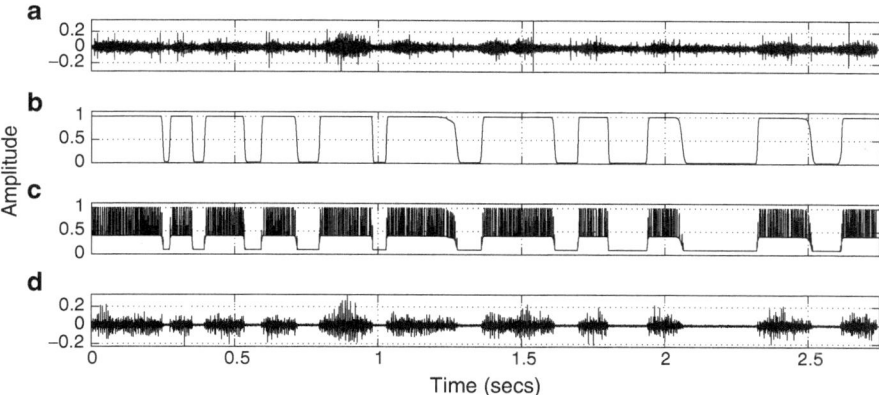

Fig. 4.7 Overall temporal processing (**a**) degraded LP residual, (**b**) gross weight function, (**c**) final weight function, and (**d**) enhanced LP residual

corresponds to the noisy LP residual shown in Fig. 4.7a is given in Fig. 4.7b. Figure 4.7c indicate the final weight function consisting of combination of gross and fine weight function. Figure 4.7d shows the enhanced LP residual after applying the final weight function. From the enhanced LP residual, it is observed that LP residual is enhanced at both gross level and final level.

The temporally processed speech sounds to be perceptually enhanced. This is mainly due to the enhancement of speech-specific regions in the noisy speech signal. This includes high SNR regions at gross level and regions around the instants of significant excitation. This is achieved by multiplying the LP residual of the noisy speech signal by the weight function. Even though the speech-specific regions are emphasized in the temporally processed speech, the noise suppression is minimal mainly due to the use of all-pole filters derived from the noisy speech. To further improve the enhancement level, the speech-specific features corresponding to the all-pole filter are enhanced subjected to spectral processing. To improve the vocal tract response characteristics at the spectral level, the spectral processing is performed on the temporally processed speech. Spectral enhancement is performed using conventional spectral subtraction [106] or minimum mean square error and short time spectral amplitude (MMSE-STSA) [108]-based methods. Spectral subtraction-based speech enhancement is performed by subtracting the average magnitude of the noise spectrum from the spectrum of the noisy speech [106]. In this method noise is assumed to be uncorrelated and additive to the speech signal. The noise estimation is obtained based on the assumption that the noise is locally stationary, so that the noise characteristics computed during the speech pauses are a good approximation to the noise characteristics.

$$\left|\widehat{S(k)}\right| = |Y(k)| - \left|\widehat{D(k)}\right|, \qquad (4.2)$$

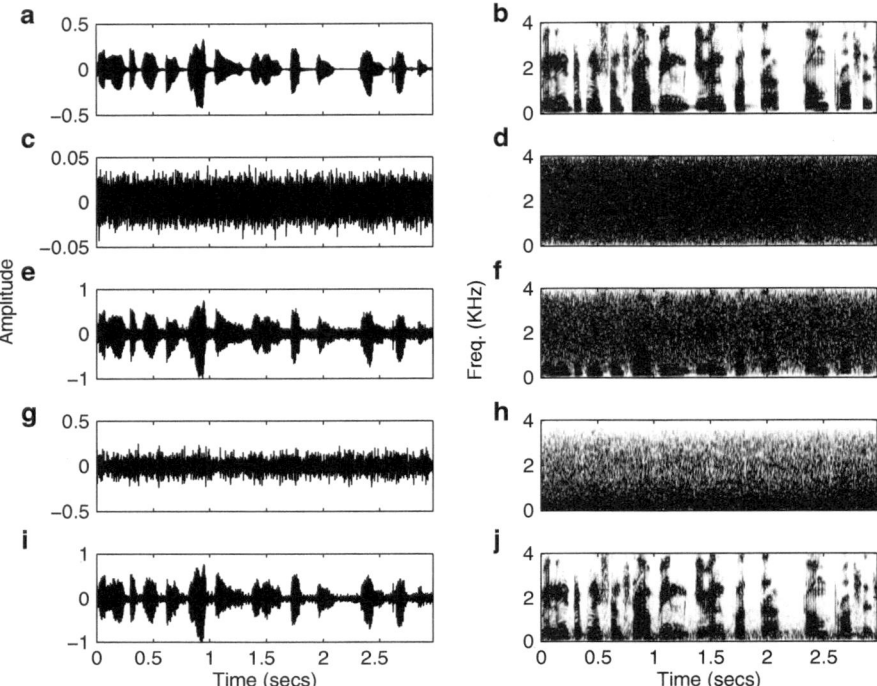

Fig. 4.8 (**a,b**) clean speech and its spectrogram; (**c,d**) white noise and its spectrogram; (**e,f**) white noise speech (SNR 10 dB) and its spectrogram; (**g,h**) vehicle noise and its spectrogram; and (**i,j**) vehicle noise speech (SNR 10 dB) and its spectrogram

where $\hat{D}(k)$ is the average magnitude of the noise spectrum, $Y(k)$ is the spectrum of noisy speech signal, and $\hat{S}(k)$ is the estimated enhanced speech signal spectrum. The MMSE-STSA for speech enhancement aims to minimize the mean square error between the short time spectral magnitude of the clean and enhanced speech signal [108]. This method assumes that each of the Fourier expansion coefficients of the speech and of the noise process can be modeled as independent, zero-mean and Gaussian random variables.

The temporal and spectral details of clean speech, noisy speech, and enhanced speech by the proposed temporal and spectral processing method are shown in Figs. 4.8 and 4.9. Figure 4.8 shows the time and frequency domain details of clean speech, noise segments, and noisy speech signals. From spectrogram plot of Fig. 4.8d it is observed that, in the case of white noise, noise is present at all frequencies (0–4 KHz) in the uniform manner. From Fig. 4.8h, it indicates that the vehicle noise is dominant at low frequencies (i.e., less than 1 KHz), and at higher frequencies (i.e., beyond 1 KHz) its effect is less. Therefore, addition of the noises to clean speech affects more in case of white noise compared to vehicle noise. This is clearly observed from the spectrograms of noisy speech (see Fig. 4.8f, j).

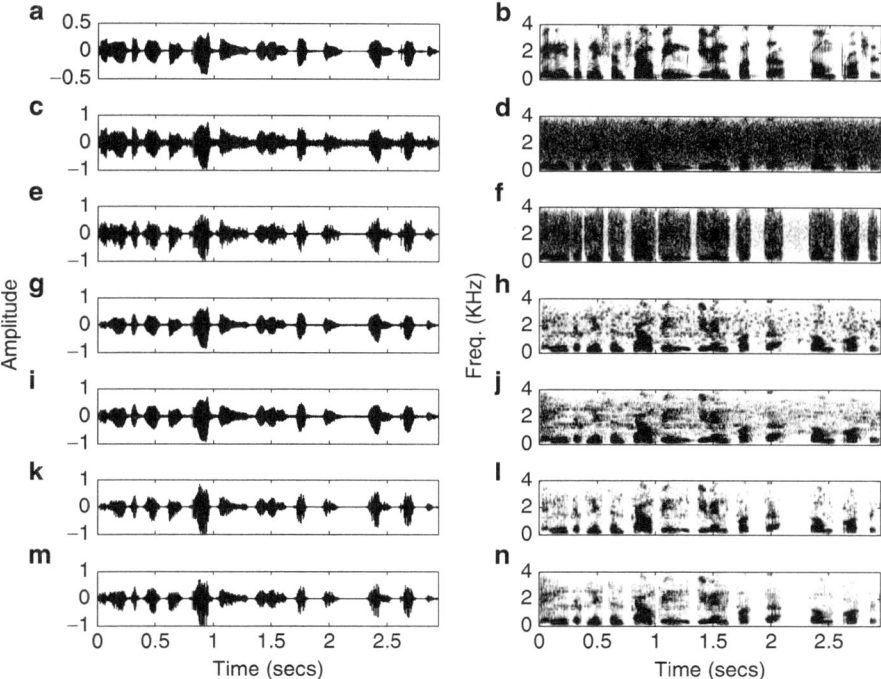

Fig. 4.9 Noisy speech enhancement through combined TSP: (**a,b**) clean speech and its spectrogram; (**c,d**) white noisy (SNR of 10 dB) speech and its spectrogram; (**e,f**) white noisy speech processed by temporal processing (TP) and its spectrogram; (**g,h**) white noisy speech processed by spectral subtraction (SS) method and its spectrogram; (**i,j**) white noisy speech processed by MMSE method and its spectrogram; (**k,l**) white noisy speech processed by combining TP and SS methods, and its spectrogram; and (**m,n**) white noisy speech processed by combining TP and MMSE methods, and its spectrogram

Figure 4.9 shows noisy and enhanced speech signals and their respective spectrograms. In this case noisy speech is derived by adding the clean speech with white noise (SNR of 10 dB). By applying the temporal enhancement, speech regions are enhanced by deemphasizing the nonspeech regions. It is observed from Fig. 4.9f that, since speech regions are preserved during temporal processing, the noise also remains along with speech components, whereas nonspeech regions are completely cleaned up because of deemphasizing the nonspeech regions. Spectral processing methods attempt to remove the noise spectral components from noisy speech. Therefore, the spectrograms of the spectrally enhanced speech (see Fig. 4.9h, j) indicate the enhanced noisy speech. In combined temporal and spectral processing methods the merits of individual methods are combined, and hence in Fig. 4.9l, n we can observe the better enhancement compared to individual methods.

Table 4.11 Overall CV recognition using proposed CV recognition method under background noise cases using different preprocessing techniques. DEG, SS, MMSE, TP, TSP1, and TSP2 refer to degraded speech, multi-band spectral subtraction, MMSE-STSA estimator, temporal processing, combined temporal and multi-band spectral subtraction and combined temporal and MMSE-STSA estimator, respectively

Enhancement method	Recognition performance (%)				
	White noise (SNR levels in dB)				
	0 dB	5 dB	10 dB	20 dB	30 dB
DEG	4.62	8.92	15.44	37.63	44.62
MMSE	26.81	32.20	38.42	46.06	54.76
TP	7.33	10.52	22.66	41.32	47.41
TSP1	**27.46**	**36.39**	**40.43**	**48.38**	**57.16**
TSP2	**29.72**	**35.31**	**41.10**	**49.12**	**57.71**
	Vehicle noise				
DEG	26.07	34.84	43.59	54.46	60.81
SS	34.23	41.34	47.81	56.20	62.21
MMSE	36.07	42.59	49.75	58.31	62.08
TP	26.08	35.12	44.57	55.12	60.88
TSP1	**37.24**	**45.46**	**52.39**	**59.24**	**63.06**
TSP2	**38.59**	**47.52**	**54.20**	**61.72**	**63.91**

4.6.2 CV Units Recognition Under Different Background Noise Cases Using Temporal and Spectral Preprocessing Techniques

Effectiveness of preprocessing techniques on CV recognition performance in noisy environments is evaluated using CV units from Telugu broadcast corpus [112]. Framework described in Sect. 4.2.3 is considered to carry out this study. Table 4.11 shows the overall CV recognition performance using proposed CV recognition method under different background noise cases using temporal and spectral preprocessing techniques. In Table 4.11 abbreviations DEG, SS, MMSE, TP, TSP1, and TSP2 refer to degraded speech, multi-band spectral subtraction, MMSE-STSA estimator, temporal processing, combined temporal and multi-band spectral subtraction, and combined temporal and MMSE-STSA estimator, respectively. From Table 4.11, it is observed that spectral processing methods provided much better improvement in the recognition performance compared to temporal processing method. This is because spectral methods enhance the noisy speech by filtering out the noise spectrum. Hence, the enhanced spectrum mostly contains speech characteristics, whereas temporal processing methods enhance the speech by processing high SNR speech regions and attenuate all other regions. With this effect, the noisy speech is perceptually enhanced, but the presence of noise within speech regions will significantly degrade the performance. Hence, in our study we have observed an improvement of 2–7 % and 4–22 % in the recognition performance using temporal method and spectral methods, respectively. From the results it is also evident that combined temporal and spectral processing techniques have shown better performance compared to individual enhancement methods. Because merits of both temporal and spectral methods are combined in combined TSP.

4.7 Summary

The aim of this chapter is to analyze the performance of CV recognition system in the presence of speech coding and background noise conditions. We have proposed a two-stage hybrid approach for developing the robust CV recognition system. In the first stage of the proposed CV recognition approach vowel was recognized and in the second stage consonant was recognized. Proposed CV recognition approach uses the combination of complimentary evidences from SVM and HMM to improve recognition performance. VOP plays a crucial role in the proposed two-stage CV recognition. Therefore, in this work we used robust and accurate VOP detection methods proposed in Chap. 3. Performance of the proposed CV recognition method is evaluated using CV utterances from isolated CV database and Telugu broadcast database. Performance of the proposed CV recognition approach is compared with single- and two-stage SVM, HMM, and single-stage SVM + HMM-based systems. The proposed CV recognition approach has shown significant improvement compared to other approaches, in the presence of clean, coded, and noisy conditions. The improved performance of the proposed method may be due to the following reasons: (1) combination of complimentary evidences from HMM and SVM and (2) accurate VOP detection method for selection of different features.

The effect of speech coding on recognition of CV units is studied for GSM FR, GSM EFR, CELP, and MELP coding techniques. From the results it is observed that the effect of GSM coders on CV recognition performance is almost minimized by using the proposed method under matched condition. Performance is significantly improved by using proposed method in case of CELP and MELP coders. Further, we studied the effect of background noise on CV recognition performance at different SNR levels. From this study, we observed that noise has severe effect on CV recognition at low SNR values. The performance of CV recognition system under background noise is improved by using combined temporal and spectral preprocessing methods. The recognition results show that combined TSP methods provide relatively higher recognition performance compared to individual methods. Future research may be carried out to study the combination of combined TSP preprocessing methods with conventional feature compensation and model adaptation methods.

Chapter 5
Spotting and Recognition of Consonant–Vowel Units from Continuous Speech

Abstract Automatic speech recognition is the process of converting speech into text. It is carried out by transforming speech signal into a sequence of symbols by using acoustic models, and converting this sequence of symbols into text by using a language model. Two approaches are commonly used for speech recognition. The first approach is based on word-level matching using word models, and then using a language model. The major drawback of this approach is to develop word models for all words of a language. In a language generally the number of words will be of order 10^5–10^6. The second approach is based on segmenting speech into subword units, and labeling them using a subword unit recognizer. The limitation of this approach lies in accurate segmentation of speech into subword units of varying durations. An approach to continuous speech recognition by spotting consonant–vowel (CV) units is presented in literature in the context of Indian languages. This approach is based on the detection of vowel onset points (VOPs) and labeling the segments around the VOPs using a CV recognizer. The major issues in this approach are accurate detection of VOPs and labeling the regions around these VOPs. In literature AANN models are used for the detection of VOPs with 30% and 6% missed and spurious rates, respectively. The performance of CV spotting and recognition using AANN models is significantly low due to inaccurate detection of VOPs. In this chapter, we propose an approach for spotting and recognition of CV units from continuous speech using accurate VOPs. Here, VOPs are determined using two-stage approach. In stage-1, VOPs are determined using the evidences from excitation source, spectral energy, and modulation spectrum of the speech segments. In stage-2, VOPs determined in stage-1 are verified and the genuine VOPs are positioned accurately using the deviation between successive epoch intervals.

In this work, VOPs are used for spotting the CV units, and the proposed two-stage CV recognition approach presented in Sect. 4.2 is used for recognition of spotted CV units. Here, VOPs are accurately determined using two-stage approach. Based on the knowledge of VOP, boundaries of vowel and consonant segments of CV

K.S. Rao and A.K. Vuppala, *Speech Processing in Mobile Environments*, SpringerBriefs in Electrical and Computer Engineering, DOI 10.1007/978-3-319-03116-3_5,
© Springer International Publishing Switzerland 2014

units are determined. After determining the boundaries of C and V, first recognition of vowel segment is carried out, and then consonant segment is recognized by performing the classification within the group of CV units corresponding to the vowel class recognized in stage-1. This chapter is organized as follows: In Sect. 5.1, proposed two-stage method for the accurate detection of VOPs is presented. Results of spotting and recognition of CV segments in continuous speech are discussed in Sect. 5.2. Spotting and recognition of CV units from coded and noisy speech are presented in Sects. 5.3 and 5.4, respectively. Summary of the findings of this chapter is presented in Sect. 5.5.

5.1 Two-Stage Approach for Detection of Vowel Onset Points

The proposed method for VOP detection is carried out in two stages. At the first stage, VOPs are detected by combining the evidence from excitation source, spectral peaks, and modulation spectrum (described in Sect. 2.1). Its accuracy of VOP detection is about 96 % within a deviation less than 40 ms, and only 45 % within 10 ms deviation [15]. But, applications like CV unit recognition require the VOP detection with high accuracy for their better performance. Therefore at the second stage, the VOPs detected in the first stage are verified (as genuine or spurious) and their locations are corrected by using the uniform epoch intervals present in the vowel regions [104].

The epoch interval corresponds to the time interval between two successive epochs. In a voiced region the epoch intervals correspond to the durations of the pitch cycles. The point at which the uniform epoch intervals start can be considered as the location of VOP [5]. Here, uniform epoch intervals correspond to the successive epoch intervals being approximately the same. Epoch intervals in a vowel region are uniform, whereas epoch intervals in an unvoiced region are nonuniform. This nature of epoch intervals in vowel regions is used in the second stage of proposed method for verification and correction of VOPs hypothesized in the first stage. Zero frequency filter method is used for determining the epoch locations and epoch intervals. In Sect. 5.1.1, sequence of steps in the proposed two-stage VOP detection method are described. Uniformity in the epoch intervals is analyzed in Sect. 5.1.2. In Sect. 5.1.3, the performance of proposed VOP detection method is compared with existing methods by using TIMIT database.

5.1.1 Sequence of Steps in the Proposed VOP
Detection Method

The proposed two-stage method for VOP detection is carried out with the following sequence of steps:

Stage-1

- Detect the VOPs by using the COMB-ESM method.

Stage-2

- Consider the epoch locations within 40 ms on both sides of each of the detected VOPs, i.e., 80 ms window around the detected VOP.
- Determine the epoch intervals within the window by calculating the difference of successive epoch locations.
- The detected VOPs are verified (as genuine or spurious) by analyzing the uniformity in the epoch intervals present in the window. The uniformity of the epoch intervals is determined by using the difference between successive epoch intervals. A threshold of 0.5 ms on the difference between successive epoch intervals is used as the upper limit for verification of a detected VOP. If a detected VOP is genuine, then there should be at least two successive epoch intervals whose difference is less than or equal to 0.5 ms.
- After deciding a detected VOP as genuine, the location of the VOP is decided based on the beginning of the uniformity in the sequence of epoch intervals. For this purpose, the values of differences between the epoch intervals are scanned from *right* to *left* side of the specified window. The location of a VOP is identified wherever the differences between the epoch intervals cross beyond the threshold. Threshold for detecting the uniformity in epoch intervals to correct the genuine VOP is determined empirically.

The proposed method for VOP detection is illustrated in Fig. 5.1 using a continuous speech utterance for the phrase *"Don't ask."* Figure 5.1a, b show the speech signal waveform and the manually marked VOPs, respectively. The VOP evidence plot and the detected VOPs obtained using the COMB-ESM method are shown in Fig. 5.1c, d, respectively. Figure 5.1e, f show the epoch locations and epoch intervals, respectively. Figure 5.1g shows the final locations of VOPs obtained using the proposed method. A spurious VOP at about sample number 8,300 (marked in bold in Fig. 5.1d) hypothesized by the COMB-ESM method is eliminated by the proposed method based on nonuniform epoch intervals around the spurious VOP (see Fig. 5.1d–f). It is also observed that the locations of genuine VOPs are corrected such that their deviations are reduced. The final VOPs detected by the proposed method are observed to be very close to the manually marked VOPs.

5.1.2 Choice of Deviation Threshold for Determining the Uniformity in the Epoch Intervals

In the proposed method, for identifying the uniformity in the epoch intervals, the difference between the successive epoch intervals is analyzed in steps of 0.125 ms (i.e., the sampling interval) from 0 to 1 ms. In a vowel region the durations of

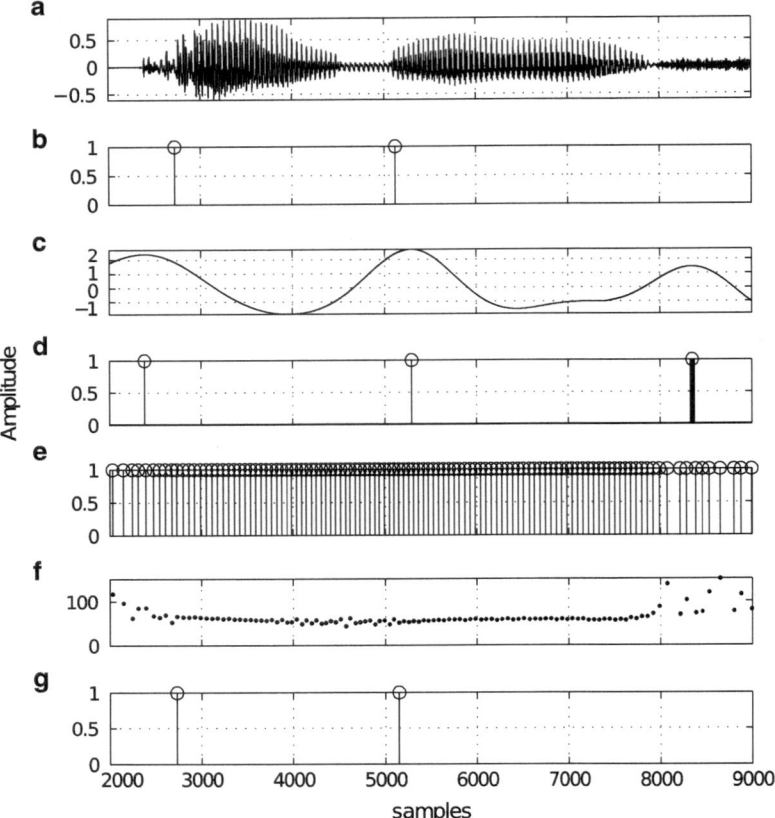

Fig. 5.1 VOP detection for a continuous speech utterance "*Don't ask*" using proposed method. (**a**) speech signal waveform, (**b**) manually marked VOPs, (**c**) COMB-ESM VOP evidence, (**d**) VOPs detected using COMB-ESM method, (**e**) epoch locations of speech signal shown in (**a**), (**f**) epoch intervals of speech signal shown in (**a**), and (**g**) VOPs detected using the proposed method

the successive pitch cycles are approximately the same, and the deviation is very small. In this work, the desired uniformity in the epoch intervals required for the elimination of spurious VOPs is analyzed by conducting the experiments on TIMIT database [77] by considering 2,407 reference VOPs. Table 5.1 shows the performance of the proposed method for detection of VOPs using different values of difference threshold used in determining the uniformity in epoch intervals. The first column shows the value of threshold on the difference between successive epoch intervals. Columns 2–4 show the number of detected VOPs, the percentage of missed VOPs, and the percentage of spurious VOPs, respectively, for each value of the threshold considered for detecting the uniformity in the epoch intervals. From Table 5.1, it is observed that for a threshold less than or equal to 0.25 ms, the percentage of missed VOPs is high and percentage of spurious VOPs is zero.

Table 5.1 Performance of the proposed VOP detection method on TIMIT database for different values of threshold on difference between successive epoch intervals

Deviation Threshold (in ms)	No. of Detected VOPs	Missed VOPs (in %)	Spurious VOPs (in %)
0	2,046	15	Nil
0.125	2,118	11.92	0
0.25	2,167	10.12	0
0.375	2,253	6.72	0.6
0.5	**2,336**	**3.98**	**0.92**
0.675	2,359	3.98	1.96
0.75	2,366	3.98	2.32
0.875	2,375	3.98	2.74
1	2,380	3.98	2.91

For a threshold higher than 0.5 ms, the percentage of spurious VOPs has increased and the missing VOP rate is 3.98%. Therefore, we have chosen the threshold as 0.5 ms for determining the uniformity in epoch intervals, and it provides the best performance of VOP detection.

5.1.3 Performance of the Proposed Two-Stage VOP Detection Method

Experiments are conducted on TIMIT database [77] for analyzing the performance of the proposed two-stage VOP detection method. The speech data of 220 sentences (120 sentences are spoken by female speakers and 100 sentences are spoken by male speakers) having 2,407 manually marked VOPs is considered. Among 2,407 VOPs, 1,013 VOPs correspond to the utterances spoken by male speakers, and the remaining 1,394 VOPs correspond to the utterances spoken by female speakers. Table 5.2 shows the accuracy in detection of VOPs using different methods. Column-1 indicates different methods considered for detecting the VOPs. Column-2 indicates the total number of VOPs detected using various methods. Columns 3–6 show the percentage of VOPs detected for 4 different values of deviation. Column-7 shows the average deviation (in ms) with respect to the manually marked VOPs. Columns 8 and 9 show the percentages of missed and spurious VOPs, respectively.

From the results, it is evident that accuracy in detection of VOP using the proposed method is observed to be superior compared to the existing methods. It is observed that about 30% more VOPs are detected within 10, 20, and 30 ms deviations using the proposed method (see columns 3–5 in last two rows of Table 5.2). Average deviation is reduced significantly in the proposed method. The percentage of spurious VOPs is also drastically reduced in the proposed method. Accuracy of the proposed method is more significant for a deviation of 20 ms.

Table 5.2 Performance of the VOP detection using excitation source (EXC), spectral peaks (SP), modulation spectrum (MOD), COMB-ESM (COMB), and proposed methods on TIMIT database

VOP detection method	Number of Detected VOPs	VOPs detected within ms (in ≈ %)				Avg. dev. (in ≈ ms)	Missing VOPs (in ≈ %)	Spu. VOPs (in ≈ %)
		10	20	30	40			
EXC	2,383	35	54	60	95	20	5	4
SP	2,384	26	46	70	94	21	6	5
MOD	2,264	36	50	74	93	18	7	2
COMB-ESM	2,386	52	60	71	96	16	4	3
Proposed	**2,336**	**84**	**95**	**96**	**96**	**7**	**4**	**1**

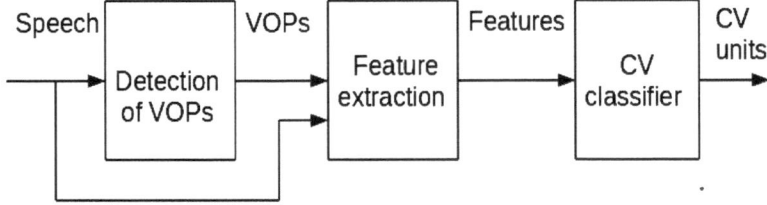

Fig. 5.2 Block diagram of CV recognition system based on spotting CV units using VOPs

The performance of the proposed method for a deviation of 10 ms is limited due to the presence of voiced consonants before the vowel onset points.

5.2 Performance of Spotting and Recognition of CV Units in Continuous Speech

The block diagram of CV recognition system based on spotting CV units from continuous speech is shown in Fig. 5.2. Initially, the VOPs are detected in continuous speech, and then the patterns extracted using the VOP locations as anchor points are used to recognize CV segments [113]. The proposed two-stage VOP detection method is used for the detection of VOPs. The CV recognition is performed using the proposed CV recognizer developed in Sect. 4.2. For evaluating the spotting and recognition of CV units from continuous speech, 500 sentences are randomly chosen from test data of Telugu broadcast news speech corpus. These 500 sentences contain 6,477 CV units. For performing the recognition of CV units from these 500 sentences, first the CV units need to be spotted. For spotting the CV units, the proposed two-stage VOP detection method is used. Performance of the proposed VOP detection method is compared with AANN [54, 78] and COMB-ESM methods [15].

In [54, 78], the AANN models are used for detection of VOPs from continuous speech. A five-layer AANN model having structure of $39L$ $60N$ $4N$ $60N$ $39L$

Table 5.3   Performance of spotting CV units using the AANN, COMB-ESM, and proposed two-stage VOP detection methods (6,477 VOPs from 500 sentences are considered)

VOP detection method	Number of matched VOPs	Number of missing VOPs	Number of spurious VOPs
AANN [54, 78]	4,436	2,041	389
COMB-ESM[15]	4,793	1,684	194
Proposed two-stage	**6,091**	**386**	**62**

is used for capturing the distributions of feature vectors. Here L refers to linear units and N refers to nonlinear units. The activation function for the nonlinear units is a hyperbolic tangent function. The AANN models are developed using 39-dimensional feature vectors (13 MFCC+ delta + delta–delta coefficients). For each CV class, two AANN models corresponding to the vowel and consonant regions are developed. For detection of VOPs in continuous speech, a feature vector is extracted from every 20 ms with 5 ms frame shift, and is given as input to the pairs of AANN models of all the CV classes. The two AANN models of each of these CV classes hypothesize the current frame as a consonant or vowel frame. Then, the current frame is labeled as consonant region or vowel region based on the majority of hypotheses from the AANN models. In this way, a sequence of labels can be obtained to a sequence of frames in continuous speech. The VOP frames are identified as those frames, at which there is a change of label from consonant to vowel [78].

Performance of the VOP detection using the AANN, COMB-ESM, and proposed two-stage VOP detection methods is shown in Table 5.3. Columns 2–4 indicate the number of matched, missed, and spurious VOPs, respectively, for different VOP detection methods. The VOPs hypothesized with a deviation less than 25 ms from actual VOPs are considered as the matched VOPs. If there are no hypothesized VOPs within 25 ms from actual VOPs, then those are considered as missed VOPs. Hypothesized VOPs beyond 25 ms around the actual VOPs are considered as spurious ones. It is seen that the proposed method for VOP detection gives an improvement of about 20% and 26% in the number of matched VOPs compared to the COMB-ESM and AANN-based VOP detection methods, respectively (see Table 5.3). The number of missing and spurious VOPs is also reduced significantly using the proposed VOP detection method. Effectiveness of the proposed VOP detection method is illustrated in Fig. 5.3 for a speech utterance "*toli jabita*" from Telugu broadcast database. The spurious VOP (third one in Fig. 5.3e) detected by the COMB-ESM method is eliminated by the proposed VOP detection method (see Fig. 5.3e, g).

After detection of VOPs using the proposed VOP detection method, patterns are extracted by using VOPs as anchor points for performing CV recognition. Patterns are extracted from different regions of a CV segment using the VOP as an anchor point for recognizing the vowel and consonant categories. Vowel region is determined by using the differences of epoch intervals starting from VOP.

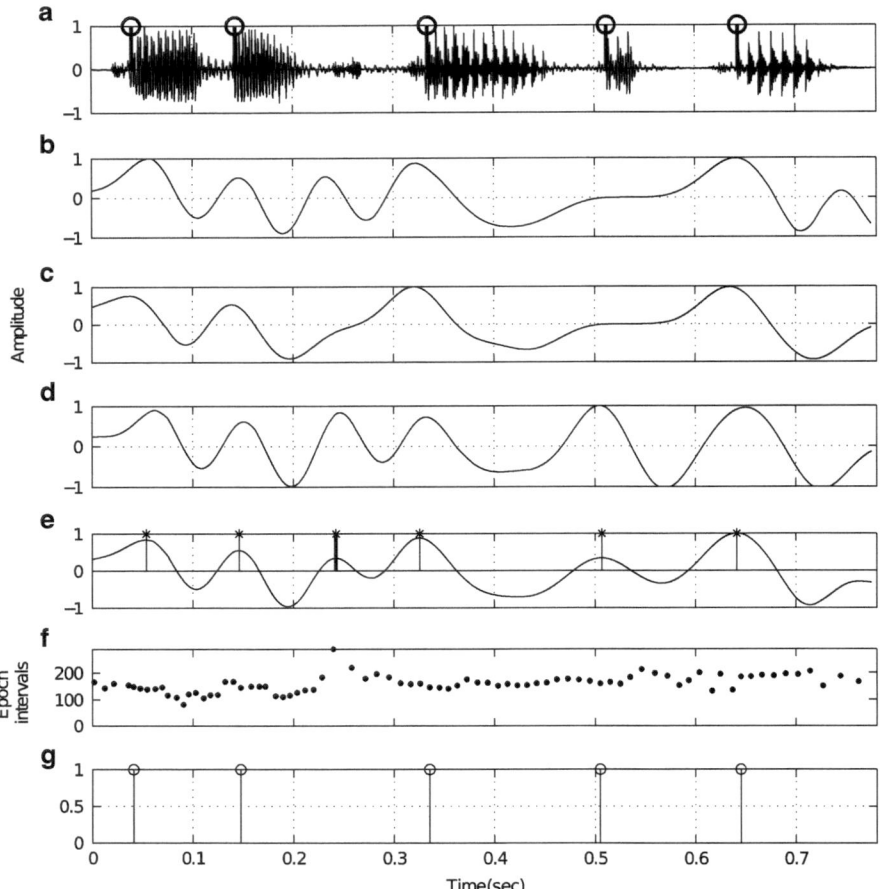

Fig. 5.3 VOP detection for a continuous speech utterance "*toli jabita*" using proposed two-stage method: (**a**) speech signal waveform with manually marked VOPs for an utterance "*toli jabita*," (**b**) VOP evidence from excitation source, (**c**) VOP evidence from spectral peaks, (**d**) VOP evidence from modulation spectrum, (**e**) VOP evidence derived by combining the evidence from excitation source, spectral peaks, and modulation spectrum [(**b**)+(**c**)+(**d**)], (**f**) epoch intervals of speech signal shown in figure (**a**), and (**g**) VOPs detected using the proposed method

Starting from the VOP to the end of uniform epoch intervals is considered as the vowel region. Uniformity in epoch intervals is determined by using same procedure described in Sect. 5.1.2. In some cases such as the following consonant of a CV unit being voiced, nasal or semivowel, uniformity in the epoch intervals continues till the following VOP. In this case, ending of the vowel region is marked using the knowledge of the consonant region of the following VOP, i.e., vowel boundary of the present CV unit is marked as the instant that is 40 ms before the following VOP. Patterns extracted from hypothesized vowel regions are used for vowel recognition.

Similarly, patterns extracted from the speech signal of 40 ms duration either side of a VOP are used for consonant recognition. CV recognition framework presented in Sect. 4.2.3 is used for this study. CV units correspond to matched VOPs detected using two-stage VOP detection method from 500 utterances are tested using the two-stage CV recognition system described in Sect. 4.2. About 63.26 % of CV segments have been correctly recognized out of 6,091 spotted CV units using matched VOPs. Overall performance of spotting and recognition of 6,477 CV units from 500 utterances is 59.48 %.

5.3 Spotting and Recognition of CV Units from Coded Speech

In this section, proposed two-stage VOP detection method is extended for spotting CV units from coded speech. From the studies, it is observed that uniformity in epoch intervals during vowel region is present even after coding. Therefore, the proposed two-stage VOP detection method can be used for accurate detection of VOPs from the coded speech. Here, two-stage VOP detection method consists of VOP detection method developed for coded speech (Sect. 3.2) in first stage for hypothesizing the VOPs. In second stage, uniformity in epoch intervals is used for removing the spurious VOPs and correcting the genuine VOPs.

Table 5.4 shows the performance of spotting CV units from coded speech using proposed two-stage VOP detection method. Spotting and recognition studies are carried out in the similar framework discussed in Sect. 5.2. From the results it is

Table 5.4 Performance of spotting CV units from coded speech using the COMB-ESM and proposed two-stage VOP detection methods (6,477 VOPs from 500 sentences are considered)

VOP detection method	Percentage of matched VOPs ($\approx$ %)	Percentage of missing VOPs ($\approx$ %)	Percentage of spurious VOPs ($\approx$ %)
Clean			
COMB-ESM	74	26	3
Proposed	**94**	**6**	**1**
GSM FR			
COMB-ESM	63	37	3
Proposed	**90**	**10**	**1**
GSM EFR			
COMB-ESM	64	36	3
Proposed	**90**	**10**	**1**
CELP			
COMB-ESM	55	45	6
Proposed	**85**	**15**	**2**
MELP			
COMB-ESM	57	43	4
Proposed	**87**	**13**	**1**

Table 5.5 Recognition performance of spotted CV units from coded speech using the proposed CV recognition system under matched condition (trained and tested under similar coded condition)

Coder	Recognition performance (%) by using matched VOPs	Overall recognition a performance (%)
PCM (clean)	63.26	59.48
GSM FR	60.92	54.83
GSM EFR	61.34	55.20
CELP	56.72	48.21
MELP	55.68	48.44

observed that performance of the proposed method is superior compared to COMB-ESM method. Spurious VOPs are reduced significantly by using proposed method under both clean and coded conditions. CV units correspond to matched VOPs are recognized using the two-stage CV recognition system discussed in Sect. 4.2.

Table 5.5 shows the recognition performance of CV units spotted from coded speech. Column-1 indicates the different coding techniques. Column-2 shows the recognition performance of spotted CV units using two-stage VOP detection method. Column-3 indicates the overall spotting and recognition of 6,477 CV units from 500 sentences. From the results, it is observed that recognition performance of CV units spotted using match VOPs is close to recognition performance of CV units using manually marked boundaries (see Tables 5.5 and 4.8).

5.4 Spotting and Recognition of CV Units from Noisy Speech

In this section, performance of spotting and recognition of CV units from noisy speech are presented. This study is carried out by adding white and vehicle noise samples from Noisex-92 [97] database to the clean speech signals at different SNR levels. Uniformity in epoch intervals during vowel region is found to be disturbed due to noise at low SNR values. Hence, it is observed that two-stage approach is not useful to enhance the performance of VOP detection. Therefore, proposed VOP detection method for noisy speech described in Sect. 3.4 is used for spotting CV units from noisy speech. Spotting and recognition studies are carried out in similar framework discussed in Sect. 5.2.

Table 5.6 shows the performance of spotting CV units from noisy speech using the COMB-ESM and proposed formant-based VOP detection methods. From the results it is observed that, performance of the formant based method is superior compared to COMB-ESM method for detecting the VOP from noisy speech. Spotted CV units corresponding to matched VOPs are recognized using the two-stage CV recognition system. Table 5.7 shows the recognition performance of CV units spotted from noisy speech. From the results, it is observed that recognition performance of CV units spotted using match VOPs is close to recognition performance of CV units using manually marked boundaries (see Tables 5.5 and 4.10).

Table 5.6 Performance of spotting CV units from noisy speech using the COMB-ESM and proposed formant-based VOP detection methods (6,477 VOPs from 500 sentences are considered)

VOP detection method	Matched VOPs(%)	Missed VOPs (%)	SPU VOPs (%)	Matched VOPs (%)	MISS VOPs (%)	SPU VOPs (%)
	White noise			Vehicle noise		
	0 dB			0 dB		
COMB-ESM	59	41	35	62	38	30
Proposed	81	19	7	82	18	6
	5 dB			5 dB		
COMB-ESM	61	39	29	63	37	26
Proposed	82	18	4	83	17	4
	10 dB			10 dB		
COMB-ESM	62	38	29	66	34	24
Proposed	82	18	3	83	17	4
	20 dB			20 dB		
COMB-ESM	66	34	16	67	33	17
Proposed	82	18	2	84	16	2

Table 5.7 Recognition performance of spotted CV units from continuous speech using proposed CV recognition system under noise

Recognition performance (%)				
White noise				
	0 dB	5 dB	10 dB	20 dB
Recog. using matched VOPs	4.81	8.96	15.31	36.92
Overall recognition	3.89	7.35	12.55	30.27
Vehicle noise				
Recog. using matched VOPs	25.19	32.62	42.22	52.37
Overall recognition	20.66	27.07	35.04	43.99

5.5 Summary

In this chapter, the performance of spotting and recognition of CV units in continuous speech has been improved by using the proposed two-stage VOP detection and two-stage CV recognition methods. The proposed VOP detection method is carried out in two stages. At the first stage, VOPs are detected by combining the evidence from excitation source, spectral peaks, and modulation spectrum. At the second stage, the detected VOPs at the first stage are verified and corrected using the uniform epoch intervals present in the vowel region. The proposed VOP detection method is shown to be effective in reducing the number of missing and spurious VOPs. There is an improvement of more than 20% in the matched VOPs by using the proposed method compared to the existing methods. Spotted CV units are recognized using the proposed two-stage CV recognition system. Studies on spotting and recognition of CV units are extended for coded and noisy speech. From the results, we observed that the proposed two-stage VOP detection approach is giving better performance, even under coding for spotting CV units. Further, methods need to be explored for minimizing the missing VOPs in case of CELP and MELP coders.

Chapter 6
Speaker Identification and Time Scale Modification Using VOPs

Abstract In this chapter, the proposed two-stage VOP detection method is used for improving the Speaker Identification (SI) performance in the presence of coding. With the help of VOPs, the crucial regions of speech segments which mainly characterize speaker-specific information are determined. Features extracted from these crucial speech segments are used for speaker identification task for improving the recognition accuracy. The accurate VOPs determined from the proposed method are also explored for nonuniform time scale modification. The proposed nonuniform time scale modification method provides high quality speech while varying speech rate. In this method, vowel regions are modified nonuniformly based on the type of vowel, and consonant and transition regions are unaltered irrespective of speaking rate. Here, vowel onset points are used to determine consonant, vowel, and transition regions.

In this chapter, two speech applications, namely speaker identification and speech rate modification based on vowel onset point (VOP) are discussed. In speaker identification task, VOPs are used to determine the important regions in speech segments where crucial speaker-specific information is present. In case of speaking rate modification, instead of modifying the entire speech signal uniformly, only vowel segments are modified according to the type of vowel and consonant and transition segments are preserved from modification. This chapter is organized as follows: In Sect. 6.2, proposed nonuniform TSM method is presented for slow down and speed up the speech. Summary of the findings of this chapter is presented in Sect. 6.3.

K.S. Rao and A.K. Vuppala, *Speech Processing in Mobile Environments*, SpringerBriefs in Electrical and Computer Engineering, DOI 10.1007/978-3-319-03116-3_6, © Springer International Publishing Switzerland 2014

6.1 Speaker Identification in the Presence of Coding Using Vowel Onset Points

In this work, initially the performance of SI system is studied with coded and cellular speech using AANN models. Later, the performance of SI system in coding environment is improved by proposing the features from steady vowel region [114,115]. Steady vowel regions are determined by using proposed method based on vowel onset points and epochs. Databases used to carry out this study are described in Sect. 6.1.1. The experimental framework used to develop SI system is presented in Sect. 6.1.2. The effect of speech coding on SI system is discussed in Sect. 6.1.3. Proposed approach for improving the performance of SI in mobile environment is presented in Sect. 6.1.4. In Sect. 6.1.5, the performance of SI systems developed using features extracted from steady vowel region and entire voiced region is compared.

6.1.1 Speech Databases

Two speech databases are recorded for analyzing the performance of SI system in mobile environment. Databases 1 and 2 were recorded simultaneously using microphone and mobile phone receiver, respectively [116]. Database-1 contains the speech corpus recorded by 100 speakers using microphone, and it is considered as clean speech. The duration of speech by each speaker is about 10 min. Coded speech corpus is simulated by passing the speech of Database-1 through standard encoder and decoder of GSM FR, GSM EFR, CELP, and MELP coders. Database-2 contains the speech corpus by 100 speakers recorded using mobile phone. For realizing the effects of speech coding and wireless channels, the database-2 is recorded at the receiving end of the mobile phone located at remote place. The sequence of steps for recording the database-2 is as follows: (1) The connection between speaker's mobile phone and destination mobile phone has to be established and (2) speaker's voice will be recorded at the destination mobile phone.

6.1.2 SI System Using AANN Models

Auto-associative neural network models are used to capture the speaker-specific information from speech signal. Speech signal contains information regarding the message, the speaker's identity, the language identity, the emotional state of the speaker, and the gender of the speaker. The uniqueness in the voice of a speaker is due to several factors such as the shape and size of the vocal tract, the dynamics of the articulators, the rate of vibration of the vocal folds, the accent imposed by the speaker, and the speaking rate. These unique characteristics are represented by

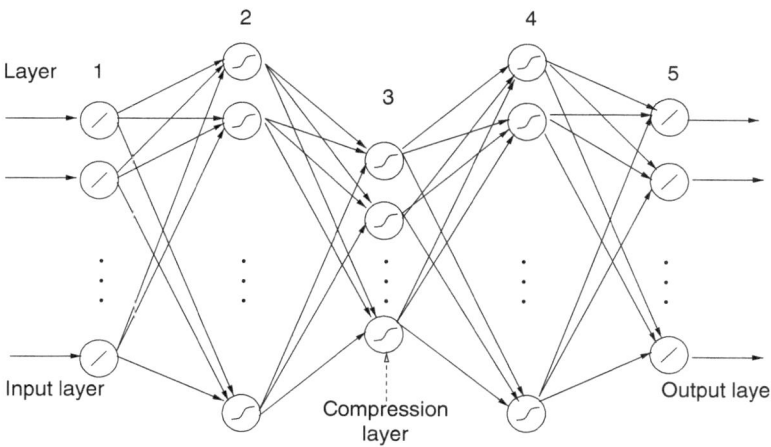

Fig. 6.1 Five-layer auto-associative neural network

speech features derived from segmental, sub-segmental, and supra-segmental levels. In this work, shape and size of the vocal tract specific to speaker is used to represent the speaker model. These speaker-specific vocal tract characteristics are represented by spectral parameters. In this work, Mel Frequency Cepstral Coefficients (MFCCs) and their velocity (delta) and acceleration (delta delta) coefficients are concatenated to form a 39-dimensional feature vectors to represent the vocal tract characteristics of a speaker. MFCC features are extracted from speech signal using frame size of 20 ms with shift of 10 ms.

Auto-associative neural networks are feedforward neural networks performing an identity mapping of the input space and are used to capture the distribution of the input data. Brief details of the AANN models are given in Appendix Sect. C.3. In this work, a five-layer AANN model [117, 118] as shown in Fig. 6.1 is used to capture the distribution of the feature vectors. The input and output (first and fifth) layers have same number of units. The second and fourth layers of the network have more units than the input layer. The third layer has fewer units than the input or output layers. All the input and output features are normalized to the range $[-1, +1]$ before presenting them to the neural network. The standard back-propagation learning algorithm is used to adjust the weights of the network to minimize the mean square error for each feature vector [117, 118].

Several neural network structures are experimented, and the final structure chosen for this study is 39L 60N 20N 60N 39L. Here, L and N indicate linear and nonlinear units. The nonlinear units use $tanh(s)$ as the activation function, where s is the activation value of that unit. SI system consists of 100 speaker models. Each speaker model is developed using auto-associative neural network. The block diagram of the SI system using AANN models is shown in Fig. 6.2. For evaluating the performance of the SI system, the feature vectors derived from the test speech utterances are given as input to all speaker models. The output of the each model is

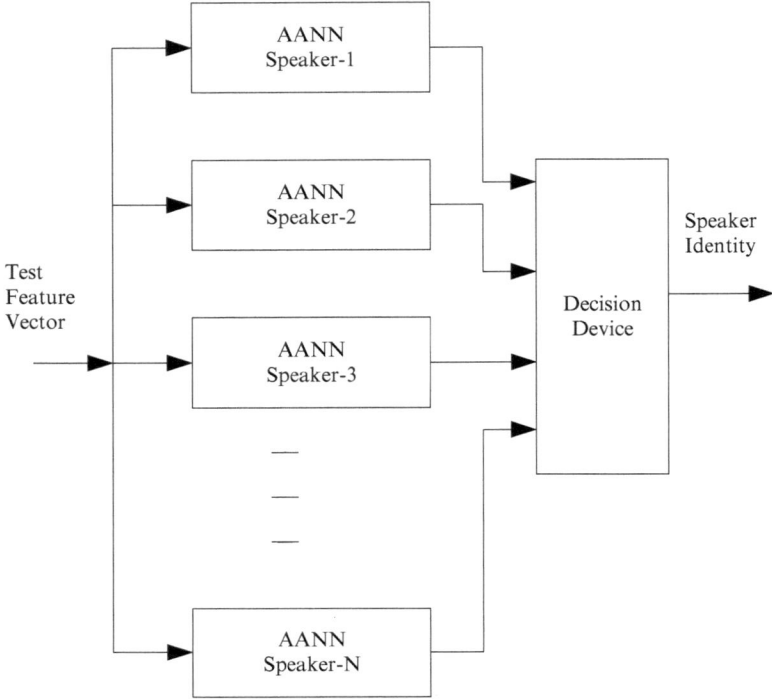

Fig. 6.2 Speaker identification system using AANN models

compared with the input to compute the normalized squared error. The normalized squared error (e) for the feature vector y is given by

$$e = ||y - o)||^2/||y||^2, \tag{6.1}$$

where "o" is the output vector given by the model. The error e is transformed into a confidence score (c) using $c = exp(-e)$. The average confidence score is calculated for each model. The identity of the speaker is decided based on the highest confidence score.

6.1.3   Effect of Speech Coding on Speaker Identification

In this section effect of speech coding on SI is analyzed by using databases described in Sect. 6.1.1. From databases 1 and 2, 5 min of speech data is used for training and 2 min of speech is used for testing. Two minutes of test speech of each speaker is divided into 24 segments, where each segment duration is about 5 s. This provides

Table 6.1 Performance of the speaker identification system under coding

Coders	Recognition performance (%)	
	Clean training	Matched training
Clean	97.12	–
GSM FR	85.22	92.33
GSM EFR	88.46	93.89
CELP	74.36	83.67
MELP	81.25	86.78
Cellular	86.46	94.42

2,400 (24 segments×100 speakers) test cases for validating the speaker models developed using databases 1 and 2.

Table 6.1 shows the performance of SI system tested with different coded and cellular speech segments. In Table 6.1, column-1 indicates the coding techniques used for testing; column-2 indicates the performance of SI models trained using clean speech and tested by speech segments derived from different coding schemes; and column-3 indicates the performance of SI models trained and tested with the corresponding speech i.e., matched condition. Results indicating that performance of the SI system is decreasing as coding rate decreases. An improvement in SI performance is observed when training and testing conditions are matched. Under matched condition, a decrement of 5, 4, 14, 11, and 3% is observed in performance of SI due to GSM FR, GSM EFR, CELP, MELP, and cellular speech, respectively. Results also indicating that SI performance is better in case of MELP compared to CELP coder, even though bit rate of MELP coder is less than CELP coder [119]. It is due to the representation of excitation signal in case of CELP coder. CELP coder uses code book to code the excitation signal approximately, whereas MELP coder uses mixed periodic and aperiodic excitations to code excitation signal effectively. So MELP preserves speaker-specific characteristics effectively compared to CELP. From the results it is evident that coding has significant effect on the performance of SI system, and hence there is a need to develop methods to improve the performance.

SI performance under coding is analyzed by computing the Pearson's correlation between clean and coded feature vectors. The Pearson's correlation coefficient r_{xy} for two variables x and y is a measure of the correlation between them [120, 121]. In this work, squared correlation r_{xy}^2 is used for analyzing the correlation between different feature vectors. Figure 6.3 shows the average squared correlation plot between MFCC features extracted from coded speech and clean speech signals. From the Fig. 6.3, it is observed that the correlation is high at lower order MFCC coefficients and gradually decreases as order of MFCC increases. Among the coders considered in this study, MFCC features derived from GSM coded speech have high correlations with the MFCCs derived from clean speech. MFCCs derived from CELP coded speech are least correlated with clean MFCCs. Similar trend is observed in the performance of SI system under coding (see Table 6.1).

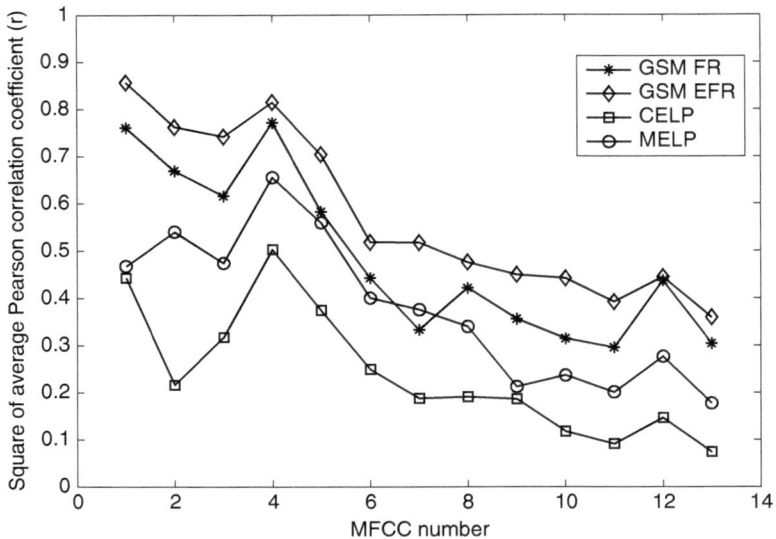

Fig. 6.3 Correlation plots between MFCC features derived for clean and coded speech

6.1.4 Proposed Speaker Identification Method

In general most of the commercial speech coders use linear predictive (LP) coefficients for coding the vocal tract information. In the LP analysis, estimation of LP coefficients is more accurate and reliable in case of steady vowel segments of speech compared to other speech segments. This is because in steady vowel speech segments the speech samples have maximum correlation, and hence the prediction error in the LP analysis is minimum. From this observation, we can hypothesize that steady vowel segments of speech are not much affected with coding, and hence the speaker-specific characteristics may be preserved in the steady vowel segments of speech even after coding. After coding, speech segments other than steady vowel region may not retain the speaker-specific characteristics, and hence the performance of SI system using the features derived from the entire speech has affected (see Table 6.1). Therefore in this work we are proposing the features only from steady vowel speech segments for carrying out the SI in case of coded and cellular speech.

Steady vowel speech segment regions are determined using the knowledge of vowel onset points and epochs similar to the proposed two-level VOP detection method presented in Sect. 5.1. Proposed method for determining steady vowel region is carried out with the following sequence of steps. (1) Detect the VOPs using proposed two-stage method described in previous chapter. (2) At each genuine hypothesized VOP, the position of the starting of steady vowel region is marked as the epoch location, where the uniform epoch intervals have started. (3) The end position of steady vowel region is marked as the epoch location, where the

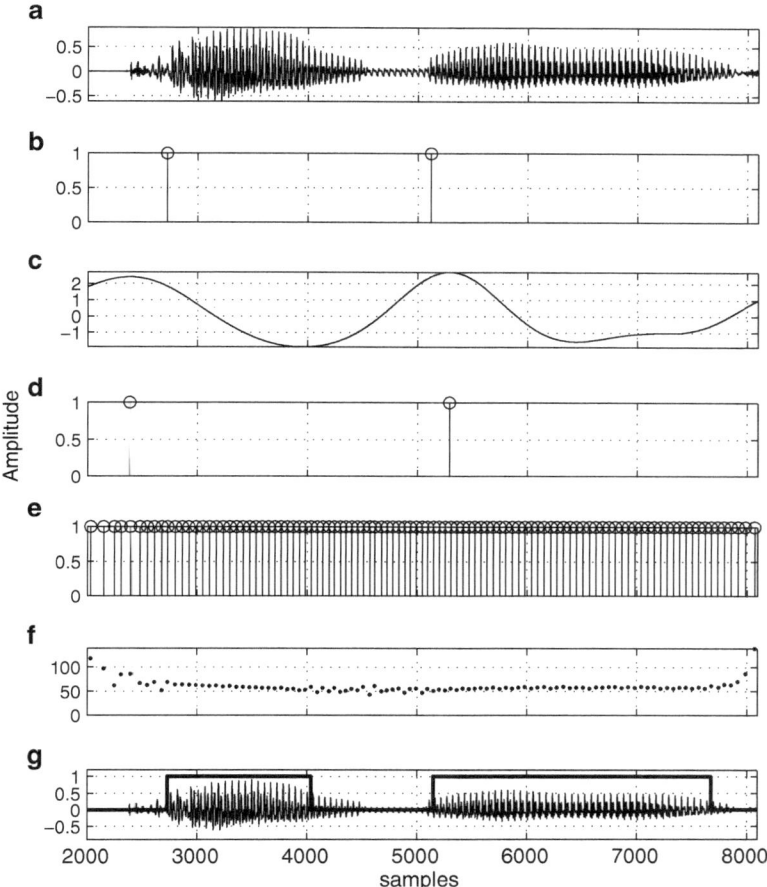

Fig. 6.4 Proposed method for determining the steady vowel regions from continuous speech. (**a**) speech signal, (**b**) manually marked VOPs, (**c**) combined VOP evidence plot, (**d**) VOPs detected using combined method, (**e**) epoch locations of speech signal shown in figure (**a**) using ZFF method, (**f**) epoch intervals of speech signal shown in figure (**a**) and (**g**) steady vowel region determined using proposed method

uniform epoch intervals have ended. Threshold for detecting the uniformity in epoch intervals is determined empirically as 0.5 ms. Method used for determining the steady vowel speech segments is illustrated by using Fig. 6.4.

In Fig. 6.4, speech signal and the manually marked reference VOPs are shown in Fig. 6.4a, b, respectively. VOP evidence plot and the detected VOPs using the COMB-ESM method are shown in Fig. 6.4c, d, respectively. Figure 6.4e, f indicate the epoch locations and epoch intervals derived from ZFF method. Figure 6.4g shows the steady vowel region (rectangular box) determined by the proposed method.

Table 6.2 Performance of SI system tested with varying coders using proposed approach

Features used for testing	Clean	GSM FR	GSM EFR	CELP	MELP	Cellular
	Recognition performance (%)					
	SI system trained with clean speech data					
Entire voiced	97.12	85.22	88.46	74.36	81.25	86.46
Steady vowel	98.03	87.27	90.22	79.54	84.56	89.33
	SI system trained with matched speech data					
Entire voiced	97.12	92.33	93.89	83.67	86.78	94.42
Steady vowel	98.03	94.53	95.14	87.44	89.24	96.37

6.1.5 Performance of the Speaker Identification System Using Features Extracted from Steady Vowel Regions

In this section, performance of SI system developed with features extracted from steady vowel region is compared with SI system developed with whole voiced speech. While extracting the steady vowel speech segments using the proposed method, it is observed that around 77 % of the voiced speech is occupied by the steady vowel speech segments. Table 6.2 shows the performance of SI system tested with different coded and cellular speech segments by using proposed method. In Table 6.2, rows 3 and 4 indicate the performance of SI system trained using clean speech and tested by different coded and cellular speech segments by using features extracted from entire voiced speech and steady vowel regions, respectively. Similarly, rows 6 and 7 indicate the performance of SI models trained and tested with the corresponding speech (i.e., matched condition). From the results shown in Table 6.2, we can observe that performance of SI system is increased by 3–4 % in case of coded and cellular speech by using proposed method. Slight improvement is observed in clean case as well by using proposed method. By using proposed method under matched condition, performance of SI system in case of GSM coded and cellular speech is comparable with clean speech case (see columns 3, 4, and 7 of Table 6.2). Significant improvement is observed in the performance of SI system in case of CELP and MELP coded speech (see columns 5 and 6 of Table 6.2) by using proposed method. From this study, it is observed that speaker-specific information present in steady vowel segments of speech is not much affected with coding.

6.2 Nonuniform Time Scale Modification Using Instants of Significant Excitation and Vowel Onset Points

The task of time scale modification (TSM) is to change the rate of speech as per requirement. In [16], authors have presented a nonuniform time scale modification method, where consonant and transition regions of speech are kept unaltered, and only vowel and pause segments are modified according to desired speaking rate [16]. In this method, vowel and pause regions are modified uniformly for fast speech as well as slow speech. In view of the production constraints of different vowel segments, all types of vowel segments may not expand or compress uniformly for different speaking rates. Therefore, in this work duration analysis is carried out for five basic vowel segments (*a, i, e, u*, and *o*) in the context of fast and slow speech [17].

In this section, the proposed nonuniform TSM method is discussed in detail. The proposed method retains the salient features of [16] and further improves the quality of speech by varying the durations of vowel segments in nonuniform manner. The goal of the proposed method is to identify different speech segments (i.e., consonant, transition, vowel, and pause) and vary the duration of each segment as per the requirements of slow and fast speaking rates of speech. Rest of the section is organized as follows: Duration analysis of five basic vowel groups for fast and slow speech is discussed in Sect. 6.2.1. Determination of different speech segments using VOPs is discussed in Sect. 6.2.2. In Sect. 6.2.3, the proposed nonuniform TSM method is described. Section 6.2.4 presents the performance of the proposed nonuniform TSM.

6.2.1 Duration Analysis of Vowels in Fast and Slow Speech

In [16], authors have conducted the duration analysis for consonant, transition, pause, and vowel segments for different speaking rates. The observations from that study are (1) durations of vowel and pause segments vary with respect to speaking rate and (2) durations of consonant and transition segments remain same for different speaking rates. In this study durations of five basic vowel categories are analyzed for fast and slow speech. Normal, fast, and slow speech data is recorded by five male and five female professional radio artists of All India Radio (AIR) station, Varanasi, India. Five Hindi sentences were chosen for recording, where the number of vowel segments in each of the five basic categories (*a, i, e, u*, and *o*) is approximately same. Speech data is collected in 5 different sessions. In each session, speech data is collected in normal, fast and slow speaking modes by each of the speakers. In this work, fast and slow speech corresponds to duration modification factors of 0.5 and 2, respectively. Altogether, about 250 sentences ($5\ sentences \times 5\ sessions \times 10\ speakers$) are collected for each of the speaking rates. Durations of the five basic vowels (*a, i, e, u*, and *o*) are analyzed manually, for normal, fast, and slow speaking rates. The variation in the durations of vowels with respect to fast and slow speaking rates is analyzed using deviation (D_i), given by $D_i = \frac{x_i - y_i}{y_i} \times 100$, where, x_i and y_i indicate durations of vowel segments for varying (fast or slow) and normal speaking rates, respectively. Table 6.3 shows the variations in durations of different vowel categories from different speakers for fast and slow speech with respect to normal speech. Column 1 indicates the speakers considered in this study. Among 10 speakers, Spk1 to Spk5 are male speakers and

Table 6.3 Percentage deviation in the durations of vowel regions under fast and slow speaking rates

| Speaker identity | % Deviation from reference duration | | | | | | | | | |
| | Slow | | | | | Fast | | | | |
	a	i	e	u	o	a	i	e	u	o
Spk1	149	79	125	81	118	44	66	46	62	54
Spk2	140	71	130	75	112	39	72	42	61	52
Spk3	142	68	126	74	118	41	63	46	57	49
Spk4	153	69	120	80	116	37	62	47	56	49
Spk5	147	73	135	71	122	45	73	44	58	51
Spk6	149	66	132	81	110	44	68	46	61	48
Spk7	140	78	127	82	109	38	65	47	59	54
Spk8	155	76	131	81	121	42	61	44	63	55
Spk9	141	70	122	70	118	43	66	47	62	53
Spk10	150	71	121	74	112	45	72	44	64	51
Average deviation	**146**	**72**	**127**	**77**	**115**	**42**	**67**	**45**	**60**	**52**

Spk6 to Spk10 are female speakers. Columns 2–6 and 7–11 show the deviations in durations of vowel segments for slow and fast speaking rates, respectively. Last row of the table indicates the average deviations in durations of vowel segments for slow and fast speaking rates. The deviations for fast and slow speaking rates are derived with reference to normal speaking rate. The +ve and −ve signed numbers in columns 2–6 and 7–11 indicate expansion and compression of vowel segment durations for slow and fast speaking rates with reference to normal speaking rate.

From the results, it is observed that the variation in durations of vowels is not same across five categories. The variability in vowel duration does also depend on speaker. While quantifying the speaker influence on the variability in vowel duration, about 6.6 % and 4 % deviations from the mean are observed for slow and fast speaking rates. From this limited study, we have not observed any specific gender influence in the variation of vowel duration. Expansion or compression of vowel depends on several factors such as (1) context (previous and following sound units), (2) position of the vowel in the word, and (3) position of the word in the phrase. In this study, we have noted the following observations: (1) among preceding and following sound units, following sound unit has more influence on the vowel duration during both expansion and compression of speech, (2) vowels at the final position of the word have higher expansion and compression factors compared to initial and middle positions, (3) vowels present in the initial words of the phrases have higher expansion factors compared to final words of the phrases, (4) vowels present in final words of the phrases have higher compression factors compared to initial and middle words during compression of speech. These observations are noted from the study on small database mentioned above, and they indicate the need for further analysis on larger database, with sufficient number of speakers and sentences to analyze the role of speaker variability, positional and contextual factors mentioned above.

For low speaking rates (slow speech), the vowels produced due to wide opening of oral cavity (i.e., *a, e,* and *o*) have expanded more compared to vowels produced due to narrow opening (i.e., *i* and *u*) of oral cavity (see column 2–6 of Table 6.3). On the other hand for high speaking rates (fast speech), vowels produced due to narrow opening of oral cavity have more compressed compared to the vowels produced with wide opening of oral cavity (see column 7–11 of Table 6.3). The amount of expansion or compression of the vowel is observed to be related to the amount of opening of oral cavity for the production of that particular vowel. The results of the above duration analysis are also seemed to be intuitive that the vowels with wide opening of oral cavity need less effort for expansion of their duration compared to their counterparts. Similarly, vowels with narrow opening of oral cavity need less effort for their compression, compared to their counterparts. The above hypothesis is more appropriate for the continuous speech having the sequence of simple CV units. In case of syllables with consonant clusters (CCV, CCCV, CVC, CCVC, and CCVCC), lot of coarticulation effects are involved in the variability of vowel durations. In Indian context, most of the languages are dominated by simple CV units [54], and hence the above hypothesis based on production constraints may be valid. Therefore, for generating the high quality speech for different speaking rates, the TSM method should modify different vowel segments by different modification factors as observed in the above analysis.

6.2.2 Determination of Different Speech Segments

In the proposed method, time scale modification is performed using instants of significant excitation. For performing nonuniform TSM, at the first step, the speech signal is divided into different speech segments, such as consonant, transition, vowel, and pause. In this work, VOPs and instants of significant excitation are used for the detection of different speech segments. VOPs are determined using proposed two-stage method (see Sect. 5.1), and ISE or epochs are extracted using ZFF method (see Sect. 3.2.1).

In Indian languages most of the characters are of the type CV or CCV (where C refers to consonant and V refers to vowel). Vowel onset point can be interpreted as the junction point between the consonant and vowel of a CV unit. The region to the left of the VOP is considered as the consonant region, and to the right of the VOP as the vowel region. In the vowel portion, a small region following the VOP is treated as transition region. After determining the vowel onset point, 30 ms to the left of the VOP is marked as the consonant region, and 30 ms to the right of the VOP is marked as the transition region. Here, we have chosen 30 ms for marking consonant and transition segments based on our previous studies [16]. The beginning of the steady vowel region is marked as 30 ms after the VOP (i.e., end point of transition region). In the proposed TSM method, the durations of consonant and transition regions are unaltered and vowel and pause regions are modified.

6.2.3 Proposed Method for Time Scale Modification

In the proposed time scale modification (TSM) method vowel segments will be modified by different modification factors based on their category, and pause segments are modified by uniform manner based on expansion or compression of speech. Consonant and transition segments of speech are unaltered during proposed TSM. For identifying various speech regions, VOPs are used as anchor points. The performance of the proposed TSM mainly depends on the accuracy of determining different speech regions such as consonant, transition, vowel, and pause. With accurate VOP locations, consonant and transition regions can be marked accurately. In this study, consonant and transition regions are marked as 30 ms speech segments, before and after each VOP, respectively. Vowel regions are marked by using VOPs and epoch intervals. Beginning of the vowel can be marked by using VOP, with an instant placed 30 ms after the VOP. End point of the vowel can be marked by using the epoch intervals and the location of the following VOP. If the vowel is followed by an unvoiced or pause segment, then end point of the vowel can be marked by using the uniformity of successive epoch intervals. As long as vowel segment is present, successive epoch intervals are uniform, and the moment vowel terminates with pause or unvoiced segment, successive epoch intervals are nonuniform. The end point of the vowel is marked as the epoch location, where uniformity in the successive epoch intervals breaks down. If the vowel is followed by voiced speech segment, uniformity of successive epoch intervals continues till the next VOP. In this case end point of the vowel is marked as an instant 30 ms before the next VOP. After marking vowel, consonant, and transition segments, the remaining segments of speech are treated as pause segments. During pause segments, the energy is very less and the epoch intervals are completely random.

After identifying different speech segments (i.e., vowel, pause, consonant, and transition), vowel classification is performed for each of the vowel segments. In this study, we have considered only five basic vowel classes (a, i, e, u, and o). Three state hidden Markov models (HMMs) are used for the classification of vowel segments. Mel-frequency cepstral coefficients are used as feature vectors. Performance of the vowel classification system is observed to be more than 97 %. After identifying the vowel category, the duration of the vowel is modified according to the observations noted in Sect. 6.2 (see Table 6.3). For generating slow speech, vowels a, i, e, u, and o need to modify by 2.46, 1.72, 2.27, 1.77, and 2.15 modification factors, respectively, and for fast speech the modification factors will be 0.42, 0.67, 0.45, 0.60, and 0.52. The modification factors mentioned above correspond to slow and fast speech, where the original duration of speech is approximately doubled and halved, respectively. For the modification of speech by desired modification factor, the vowel modification factors can be derived by scaling the above factors as per the requirement. With the proposed nonuniform TSM method, slight deviation is observed between the expected duration and the actual duration obtained after the modification. This deviation is due to modification of different speech segments by different factors and also depends on the frequency of occurrence of specific vowel

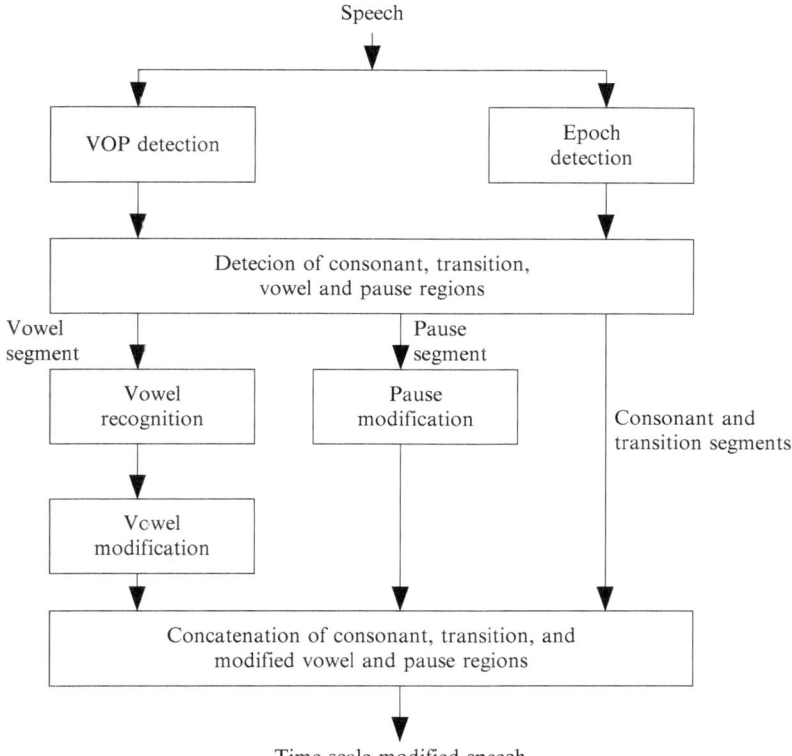

Fig. 6.5 Block diagram of proposed nonuniform time scale modification system

segments. In the present study, the deviation between actual and expected durations is found to be less than 3 %, with respect to expected duration. The block diagram of the proposed nonuniform TSM system is shown in Fig. 6.5.

In this work, epoch-based method is used for modification of vowel and pause durations [59, 63]. The basic steps in epoch-based method are as follows: (1) Derive the epoch sequence for the given vowel or pause segment. (2) Derive the new epoch sequence according to the modification factor. (3) Modify the speech signal according to the new epoch sequence. Here, each epoch is associated with time, epoch interval and speech signal correspond to that epoch interval. For time scale expansion, the new epoch sequence consists of insertion of some new epochs, whereas for time scale compression the new epoch sequence consists of deletion of some epochs from the original epoch sequence.

Generation of new epoch sequence for time scale modification is illustrated in Fig. 6.6 for a duration increase by $\beta = 1.5$ times, and in Fig. 6.7 for a duration decrease by $\beta = 0.75$ times. For generating the desired epoch interval plot for time scale modification, the original epoch interval plot (solid lines in Figs. 6.6 and 6.7)

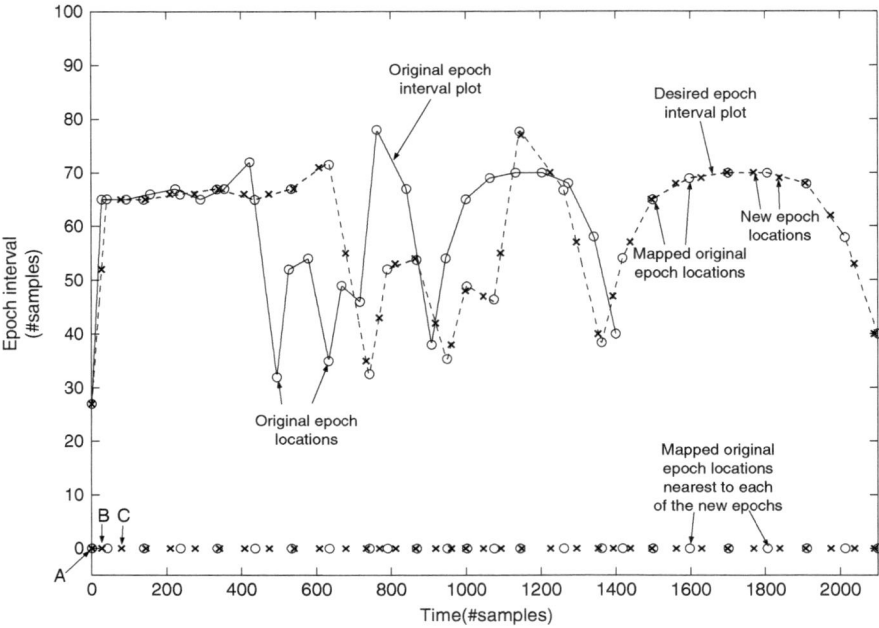

Fig. 6.6 Generation of new sequence of epochs for the modification of duration by a factor 1.5

is resampled according to the desired modification factor. The desired epoch interval plot is shown by the dotted curve. The modified (new) epoch sequence is generated as follows. Starting with the point A in Fig. 6.6, the epoch interval value is obtained from the dotted curve, and it is used to determine the next epoch instant B. The value of the next epoch interval at B is obtained from the dotted curve, and this value is used to mark the next new epoch C. The new epochs generated by this process are marked as 'x' along the x-axis in Fig. 6.6. The new epochs are also marked ('x') on the desired epoch interval plot along with the mapped original epochs ('o'). Those mapped original epochs nearest to the new epochs are shown along the x-axis by circles ('o'). In a similar manner, the new epochs are generated for the case of decrease of duration and are shown in Fig. 6.7. Here, for describing the epoch-based method, we have considered speech segment of 1,400 samples, with two voiced segments (1–400 and 900–1,400 samples) and a pause segment (400–900 samples) (see Figs. 6.6 and 6.7). It can be noted from Figs. 6.6 and 6.7 that epoch-based method does not discriminate voiced and pause segments for performing TSM.

After obtaining the modified epoch sequence, the next step is to generate the speech signal according to the modified epoch sequence. For this, the original epoch (represented by a 'o') closest to the modified epoch ('x') is determined from the sequence of 'o' and 'x' along the desired epoch interval curve (dotted curves in each of the Figs. 6.6–6.7). As mentioned earlier, with each original epoch, i.e., the circles ('o') in the plots, there is an associated speech signal of length equal to the

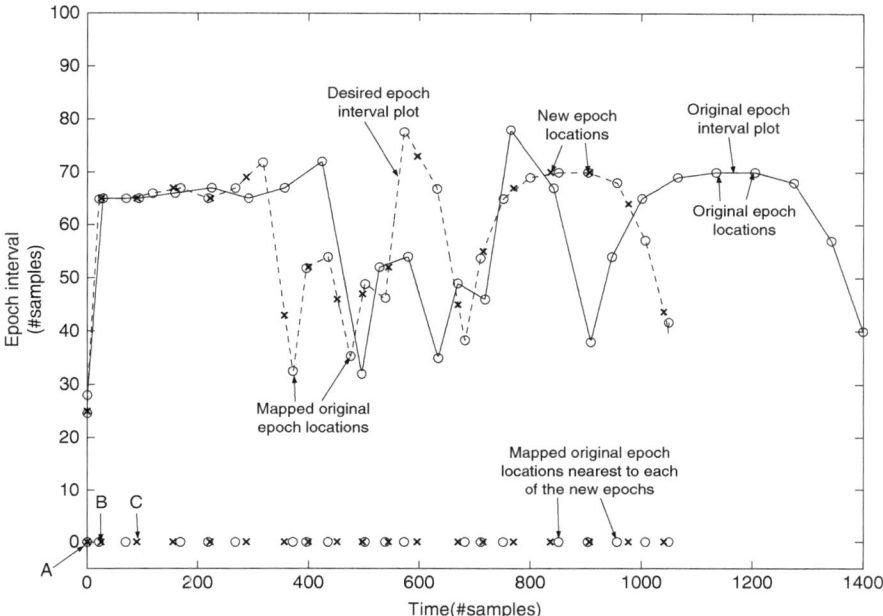

Fig. 6.7 Generation of new sequence of epochs for the modification of duration by a factor 0.75

Table 6.4 Steps for the proposed time scale modification method

1.	Derive the epochs from speech signal using zero frequency filter method
2.	Derive the vowel onset points from speech signal using multiple evidences derived from Hilbert envelope of LP residual, spectral peaks, and modulation spectrum
3.	Improve the accuracy of derived VOPs using the knowledge of epoch intervals present in vowel region
4.	Mark the consonant and transition regions as 30 ms speech segments before and after the location of each VOP
5.	Mark the vowel and pause segments using VOPs and epoch intervals
6.	After detecting the vowel segments, classify them using hidden Markov models
7.	Vowel segments are modified by specific modification factors, depending on their category by using epoch-based method
8.	Pause segments are modified using epoch-based method
9.	Concatenate the unmodified consonant and transition segments and modified vowel and pause segments in the sequence present in the original speech signal

value of the original epoch interval for that epoch. The speech samples are placed starting from the corresponding new epoch. In this way, speech signal is generated for the new epoch sequence corresponding to slow or fast speech. The sequence of steps in the proposed TSM method is given in Table 6.4.

Table 6.5 Ranking used for judging the quality and perceptual distortion of the speech signal modified by different modification factors

Rating	Speech quality	Level of perceptual distortion
1.	Unsatisfactory	Very annoying and objectionable
2.	Poor	Annoying but not objectionable
3.	Fair	Perceptible and slightly annoying
4.	Good	Just perceptible but not annoying
5.	Excellent	Imperceptible

6.2.4   Evaluation of the Proposed Nonuniform Time Scale Modification Method

Performance of the proposed epoch-based time scale modification method is compared with the well-known time-domain pitch synchronous overlap-and-add (TD-PSOLA) method using perceptual evaluation. The reason for choosing TD-PSOLA-based TSM method for comparison is that TD-PSOLA also performs time scale modification directly on speech signal using pitch synchronous markers similar to proposed method. In this work, performance of the proposed epoch-based method and TD-PSOLA-based method is analyzed by carrying out TSM in following manner: (1) uniform TSM (all speech segments are modified uniformly according to desired modification factor), (2) nonuniform TSM-1 (vowel and pause segments are modified uniformly as per the desired modification factor, and consonant and transition regions are preserved from modification [16]), and (3) nonuniform TSM-2 (vowel segments are modified by specific modification factors based on their category, pause segments are modified by uniform manner, and consonant and transition regions are preserved from modification). The main aim of these studies is to analyze the effectiveness of the proposed nonuniform TSM compared to uniform TSM using well-known TSM methods.

Perceptual evaluation was carried out by conducting subjective listening tests with 30 research scholars in the age group of 25–40 years. The subjects have sufficient speech knowledge for proper assessment of the speech signals, as all of them have taken a full semester course on speech technology. Five sentences were chosen to perform the test. Speech signals were derived for the fast and slow speaking rates using the modification factors of 0.5, 0.75, 1.25, 1.5, 1.75, and 2. Epoch and TD-PSOLA methods are used to modify the speech signals by the above-mentioned modification factors using uniform and nonuniform modification schemes. The tests were conducted in the laboratory environment by playing the speech signals through headphones. Test speech samples are played to subjects in a random fashion, by hiding the details of method of TSM. In the test, the subjects were asked to judge the perceptual distortion and quality of the speech for fast and slow speech samples. Subjects were asked to assess the quality and perceptual distortion on a 5-point scale for each of the sentences obtained by both the methods. The 5-point scale for representing the quality of speech and the distortion level is given in Table 6.5 [122].

The Mean Opinion Scores (MOSs) for fast and slow speech samples derived from epoch and TD-PSOLA methods using uniform and nonuniform TSM schemes are given in Table 6.6. The level of confidence (LOC) is computed for the difference of each pair of MOSs [123] derived from epoch and TD-PSOLA methods under different modification schemes. Results of perceptual evaluation show that performance of epoch-based method is slightly better compared to TD-PSOLA method under uniform and nonuniform modification schemes. This may be due to retaining the original speech samples in each epoch interval (pitch cycle) as it is during TSM in epoch-based method, whereas in case of TD-PSOLA, TSM is performed by overlapping and adding the adjacent speech segments. During this process sometimes phase mismatch between segments may result slight distortion. But, the difference in the MOSs between the methods is not statistically significant.

From the observation of MOSs between uniform and nonuniform TSM schemes, the performance of nonuniform schemes is found to be superior compared to uniform TSM. Among nonuniform TSM schemes, performance of the proposed nonuniform TSM (nonuniform TSM-2) is better compared to nonuniform TSM-1. At lower modification rates both uniform and nonuniform schemes produce good quality speech. When the modification factors are less than 0.75 for compression, and more than 1.5 for expansion, nonuniform schemes produce better quality of speech compared to uniform TSM scheme. At the extreme cases (modification factor 2 for expansion and 0.5 for compression) proposed nonuniform TSM scheme is better compared to nonuniform TSM-1 scheme. The above observations are also verified by using hypothesis testing.

Table 6.7 shows the level of confidence for the significance of difference between pairs of MOSs for the speech samples derived from epoch-based method with uniform and nonuniform TSM schemes. Columns 2 and 3 indicate LOC values for the pairs of MOSs obtained from uniform TSM and proposed nonuniform TSM and nonuniform TSM-1 and proposed nonuniform TSM methods, respectively. It is observed that LOC values in column 2 are highly compared to LOC values in column 3. This shows that the improvement in the performance of the proposed nonuniform TSM is more significant with respect to uniform TSM compared to nonuniform TSM-1. On the whole, from the overall observation we can conclude that the proposed nonuniform TSM method gives intelligible and better quality speech. The superior performance of the proposed method is due to nonuniform modification of different speech segments and accurate detection of various speech segments with the help of ISE and VOPs.

6.3 Summary

In this chapter, we have addressed the problem of speaker identification in mobile environment, paying attention to distortion due to speech coding. The effect of speech coding on the performance of speaker identification system is studied by using AANN models. The performance of speaker identification system was

Table 6.6 Mean opinion scores and confidence values for fast and slow speech signals derived using epoch and TD-PSOLA methods

Speaking rate (modification factor)	Uniform TSM			Nonuniform TSM-1			Nonuniform TSM-2		
	Mean opinion score		LOC in (%)	Mean opinion score		LOC in (%)	Mean opinion score		LOC in (%)
	TD-PSOLA method	Epoch method	between the pair of MOSs	TD-PSOLA method	Epoch method	between the pair of MOSs	TD-PSOLA method	Epoch method	between the pair of MOSs
Fast (0.5)	3.18	3.51	> 99.0	3.59	3.86	> 97.5	3.91	4.12	> 97.5
Fast (0.75)	4.26	4.32	< 90.0	4.36	4.43	< 90.0	4.47	4.52	< 90.0
Slow (1.25)	4.61	4.65	< 90.0	4.68	4.72	< 90.0	4.73	4.78	< 90.0
Slow (1.5)	4.32	4.48	> 95.0	4.49	4.62	> 95.0	4.62	4.75	> 95.0
Slow (1.75)	4.03	4.27	> 97.5	4.31	4.49	> 97.5	4.50	4.69	> 97.5
Slow (2.0)	3.91	4.16	> 97.5	4.19	4.36	> 97.5	4.36	4.58	< 99.0

Table 6.7 Confidence values for fast and slow speech signals derived by epoch-based method using uniform and nonuniform TSM schemes

Speaking rate (modification factor)	Level of confidence in (%) for the significance of difference in MOSs	
	Uniform TSM	Nonuniform TSM-1
Fast (0.5)	> 99.5	> 99.5
Fast (0.75)	> 97.5	> 95.0
Slow (1.25)	> 95.0	< 90.0
Slow (1.5)	> 97.5	> 95.0
Slow (1.75)	> 99.5	> 97.5
Slow (2.0)	> 99.5	< 99.0

evaluated using microphone, coded and cell phone speech data. From the results it is evident that performance of speaker identification system is decreased with speech coding, the performance is observed to be improving in matched condition.

Proposed speaker identification system developed by the features from steady vowel speech segments has shown better performance compared to features from the entire voiced speech. This is mainly due to the presence of crucial speaker-specific information in the steady vowel segments of speech even after coding. Steady vowel segments were determined by using the knowledge of vowel onset points and epochs. From the results we observed that performance of speaker identification system under GSM coders and cellular speech is comparable with clean speech by using proposed method. Significant improvement in SI is observed in case of CELP and MELP coders by using proposed method.

A nonuniform time scale modification method is proposed for producing more intelligible and natural speech for different speaking rates. In this method, speech signal is processed directly using instants of significant excitation for maintaining the original pitch variations as well as varying the durations of different speech segments in nonuniform manner. The knowledge of vowel onset points and instants of significant excitation is used for detecting different speech segments as well as modifying their durations by different modification factors. The effectiveness of the proposed method depends mainly on the accuracy in detecting the instants of significant excitation and the locations of the vowel onset points, because the detection and modification of different speech segments are carried out using the epochs and VOPs as anchor points. Subjective tests indicate that the performance of the proposed nonuniform TSM method is superior compared to existing nonuniform and uniform TSM methods. By exploiting the duration characteristics of various speech units with respect to their linguistic context and production constraints at different speaking rates may improve the quality of speech further, for different speaking rates.

Chapter 7
Summary and Conclusions

Abstract This book discusses some important issues in speech processing in the context of mobile environment. The major challenges in speech processing in mobile environment are: varying background conditions, speech coding and transmission channel errors. This book suggests signal processing methods to determine some crucial events in speech, which are robust to above-said adverse conditions. In this work, authors have proposed vowel onset points (VOPs) as crucial events in speech, which are robust to speech coding and background noisy environments. By using VOPs as anchor points, speech signals are processed in the presence of coding and noisy environments for developing the speech systems such as speech recognition, speaker recognition, and speaking rate modification. From the results, it is grossly observed that the performance of developed speech systems is superior compared to systems developed without using the knowledge of VOPs. This chapter summarizes the findings of the present work, highlights the major contributions, and flashlights on the directions for future work.

7.1 Summary of the Present Work

- *VOP detection from coded speech*: Existing VOP detection methods are suffering with low detection accuracy [15,93,124]. Therefore, we have proposed a method for accurate detection of VOP for clean and coded speech [87]. The proposed VOP detection method is based on the spectral energy in 500–2,500 Hz frequency band of the speech segments present in the glottal closure region. The reason for choosing the spectral energy in the 500–2,500 Hz band is that the vowel energy in this band is much higher than the consonant. Effect of speech coding on the performance of VOP detection was studied for GSM FR, GSM EFR, CELP, and MELP coding techniques. From this study we observed that reduction in VOP detection performance was not significant in case of GSM coders, and significant in case of CELP and MELP coders. VOP detection methods based

K.S. Rao and A.K. Vuppala, *Speech Processing in Mobile Environments*, SpringerBriefs in Electrical and Computer Engineering, DOI 10.1007/978-3-319-03116-3_7, © Springer International Publishing Switzerland 2014

on vocal tract system characteristics have performed better compared to other methods in the presence of coding. This is because speech coders preserve the vocal tract characteristics of the speech signal. From the performance evaluation, we have observed that proposed VOP detection method was superior compared to existing methods under both clean and coded conditions. It is also observed that performance of proposed VOP detection method is degrading, as bit rate of coders decreasing. But, this degradation is less in proposed VOP detection method comparing to existing methods.

- *VOP detection from noisy speech*: Performance of the VOP detection methods under background noise was studied by using white and vehicle noises at different SNR values [98]. The performance of existing VOP detection methods is significantly affected due to spurious VOPs at low SNR values. Therefore, we have proposed a VOP detection method based on spectral energy at formant frequencies of the speech segments present in glottal closure region [94]. Here, we considered spectral energy at formant frequencies instead of 500–2,500 Hz band energy, because spectral energy in 500–2,500 Hz band may not be robust under noise. Performance of the proposed method was observed to be superior compared to existing methods, and spurious VOPs were also reduced significantly. The improved performance of the proposed method is due to exploiting the high SNR characteristics of speech present at the formant frequencies in the glottal closure phase.

- *Recognition of CV units in the presence of coding*: The effect of speech coding on recognition of CV units was studied for GSM FR, GSM EFR, CELP, and MELP coding techniques [99]. From this study, it was observed that coding has significant effect on CV recognition performance. We have proposed a two-stage hybrid SVM-HMM-based approach for developing the robust CV recognition system [100]. In the first-stage of proposed CV recognition approach, vowel was recognized, and in the second-stage, consonant was recognized. At both the stages, the combination of complimentary evidences from SVM and HMM was used to enhance the recognition performance. Performance of the proposed CV recognition method was evaluated using CV utterances from isolated CV database and Telugu broadcast database. Performance of the proposed CV recognition approach was compared with existing single- and two-stage SVM, HMM-based systems. Performance of the proposed CV recognition approach has shown significant improvement compared to existing approaches in the presence of both clean and coded conditions. The improved performance of the proposed method may be due to (1) combination of complimentary evidences from HMM and SVM and (2) accurate VOP locations derived from the proposed method for selection of appropriate features for developing the CV recognition system. From the results it was observed that the effect of GSM coders on CV recognition performance was almost nullified by using the proposed method under matched condition. In the case of CELP and MELP coders, the CV recognition performance was significantly improved by using proposed method.

- *Recognition of CV units in the presence of background noise*: The effect of background noise on CV recognition was studied by using proposed CV recognition system for white and vehicle noise cases [24]. Results indicate that noise has significant effect on CV recognition performance. The proposed CV recognition approach has shown improvement in the recognition performance compared to existing methods [112]. Further, the performance of CV recognition system under background noise was improved by using combined temporal and spectral processing-based preprocessing methods. From this study, it was observed that combined TSP method provides relatively better performance compared to individual temporal or spectral methods. It is because of the fact that merits of both temporal and spectral processing methods are combined in combined TSP method.

- *Spotting and recognition of CV units from continuous speech*: The major issues in continuous speech recognition by spotting CV units are accurate detection of VOPs and labeling the regions around these VOPs. We have proposed two-stage VOP detection method for the accurate detection of VOPs [104]. At the first stage of the proposed method, VOPs were detected by combining the evidence from excitation source, spectral peaks, and modulation spectrum. At the second stage of the proposed method, the detected VOPs at the first stage were verified and corrected using the uniform epoch intervals present in the vowel region. Proposed VOP detection method has shown to be useful in reducing the number of missing and spurious VOPs significantly, compared to existing methods. The spotted CV units were recognized using a proposed two-stage CV recognition system. This study was extended to coded and noisy speech, and observed the similar result [113].

- *Speaker identification in the presence of coding using VOPs*: The effect of speech coding on the performance of speaker identification system was studied by using AANN models [116, 119]. The developed speaker identification system was evaluated using microphone, coded and cell phone speech data. From the results we have observed that coding has significant effect on SI performance. Performance of SI system was improved by using the features from the steady vowel speech segments [114, 115]. In this work, steady vowel segments were identified using VOPs.

- *Nonuniform time scale modification using instants of significant excitation and vowel onset points*: We have proposed a nonuniform time scale modification method for producing more intelligible and natural speech using accurate ISE and VOPs [17]. In the proposed TSM method, speech signal was processed directly using ISE for maintaining the original pitch variations as well as varying the durations of different speech segments in nonuniform manner. In the proposed TSM method vowel segments were modified by different modification factors based on their category, and pause segments were modified uniformly as per the desired modification factor. The modification factors associated with different vowel segments for slow and fast speech were based on their production and articulatory constraints. Consonant and transition segments of speech were unaltered during proposed TSM. Proposed two-stage VOP detection method was used for determining different segments of the speech [104]. From the subjective tests we observed that the performance of the proposed nonuniform TSM method was superior compared to existing nonuniform and uniform time scale modification methods [17,59,63].

7.2   Contributions of the Present Work

The important contributions of the research work reported in this book are as given below:

- Methods for the detection of VOPs for coded and noisy speech are proposed.
- Two-stage hybrid approach is proposed for recognition of CV units.
- Two-stage VOP detection method is proposed for spotting CV units from continuous speech.
- Combined temporal and spectral preprocessing methods are explored to improve the performance of CV recognition system under background noise.
- Improved speaker identification system in mobile environment is proposed using VOPs.
- A nonuniform time scale modification method using VOPs and ISEs is proposed for producing more intelligible and natural speech.

7.3   Directions for Future Work

- In this work, the effect of speech coding and noisy conditions is analyzed. Further, the effect of channel impairments on speech systems needs to be studied.
- Speech coding in the presence of background noise imposes an additional challenge to speech systems in mobile environments. The coding techniques designed based on speech production characteristics may affect due to noise. Therefore, the effect of coding in the presence of noise needs to be analyzed critically for building robust speech systems in mobile environments.
- In the proposed CV recognition approach, CV units are divided into subgroups based on vowel identity to improve the recognition performance of CV units. Further division of CV units based on place of articulation and manner of articulation may enhance the CV recognition performance. Hence, there is a need to explore the characteristics of place and manner of articulations of sound units.
- In this work, we analyzed the effect of speech coding and background noise conditions on consonant–vowel recognition. Further, analysis needs to be extended for automatic speech recognition in Indian languages.
- Features extracted from steady vowel region may enhance the performance of speaker identification under background noise. Hence, methods need to be explored for the determination of steady vowel region from noisy speech.

- In this work, we explored spectral (MFCC) features for building speech systems. Combination of excitation or prosodic features with spectral features may enhance the recognition performance of speech systems in adverse conditions. Hence, analysis needs to be extended by using excitation source and prosodic features.
- The effectiveness of proposed VOP detection and CV recognition methods needs to be analyzed further for new and upcoming coding standards.
- The analysis of proposed methods needs to be extended for voice over IP technology and Internet-based speech paradigm.
- The proposed nonuniform time scale modification modifies the vowel segments based on their production and articulatory constraints. The performance may be further improved by incorporating coarticulation and pause characteristics of fast and slow speech.
- In this work, different regions of speech such as consonant and transition segments are measured as fixed duration segments with respect to VOPs. Appropriate methods need to be developed for accurate detection of consonant and transition segments based on their characteristics.

Appendix A
MFCC Features

The MFCC feature extraction technique basically includes windowing the signal, applying the DFT, taking the log of the magnitude, and then warping the frequencies on a Mel scale, followed by applying the inverse DCT. The detailed description of various steps involved in the MFCC feature extraction is explained below.

1. *Pre-emphasis:* Pre-emphasis refers to filtering that emphasizes the higher frequencies. Its purpose is to balance the spectrum of voiced sounds that have a steep roll-off in the high frequency region. For voiced sounds, the glottal source has an approximately -12 dB/octave slope [101]. However, when the acoustic energy radiates from the lips, this causes a roughly $+6$ dB/octave boost to the spectrum. As a result, a speech signal when recorded with a microphone from a distance has approximately a -6 dB/octave slope downward compared to the true spectrum of the vocal tract. Therefore, pre-emphasis removes some of the glottal effects from the vocal tract parameters. The most commonly used pre-emphasis filter is given by the following transfer function:

$$H(z) = 1 - bz^{-1} \qquad (A.1)$$

where the value of b controls the slope of the filter and is usually between 0.4 and 1.0 [101].

2. *Frame blocking and windowing:* The speech signal is a slowly time-varying or quasi-stationary signal. For stable acoustic characteristics, speech needs to be examined over a sufficiently short period of time. Therefore, speech analysis must always be carried out on short segments across which the speech signal is assumed to be stationary. Short-term spectral measurements are typically carried out over 20 ms windows and advanced every 10 ms [125, 126]. Advancing the time window every 10 ms enables the temporal characteristics of individual speech sounds to be tracked and the 20 ms analysis window is usually sufficient to provide good spectral resolution of these sounds, and at the same time short enough to resolve significant temporal characteristics. The purpose of the overlapping analysis is that each speech sound of the input sequence would be

K.S. Rao and A.K. Vuppala, *Speech Processing in Mobile Environments*, SpringerBriefs in Electrical and Computer Engineering, DOI 10.1007/978-3-319-03116-3, © Springer International Publishing Switzerland 2014

approximately centered at some frame. On each frame a window is applied to taper the signal toward the frame boundaries. Generally, Hanning or Hamming windows are used [101]. This is done to enhance the harmonics, smooth the edges, and reduce the edge effect while taking the DFT on the signal.

3. *DFT spectrum:* Each windowed frame is converted into magnitude spectrum by applying DFT.

$$X(k) = \sum_{n=0}^{N-1} x(n)e^{\frac{-j2\pi nk}{N}}; \qquad 0 \le k \le N - 1 \qquad (A.2)$$

where N is the number of points used to compute the DFT.

4. *Mel-spectrum:* Mel-Spectrum is computed by passing the Fourier transformed signal through a set of band-pass filters known as mel-filter bank. A mel is a unit of measure based on the human ears perceived frequency. It does not correspond linearly to the physical frequency of the tone, as the human auditory system apparently does not perceive pitch linearly. The mel scale is approximately a linear frequency spacing below 1 kHz, and a logarithmic spacing above 1 kHz [127]. The approximation of mel from physical frequency can be expressed as

$$f_{mel} = 2595 \log_{10}\left(1 + \frac{f}{700}\right) \qquad (A.3)$$

where f denotes the physical frequency in Hz, and f_{mel} denotes the perceived frequency [125].

Filter banks can be implemented in both time domain and frequency domain. For MFCC computation, filter banks are generally implemented in frequency domain. The center frequencies of the filters are normally evenly spaced on the frequency axis. However, in order to mimic the human ears perception, the warped axis according to the nonlinear function given in Eq. (A.3) is implemented. The most commonly used filter shaper is triangular, and in some cases the Hanning filter can be found [101]. The triangular filter banks with mel-frequency warping is given in Fig. A.1.

The mel spectrum of the magnitude spectrum $X(k)$ is computed by multiplying the magnitude spectrum by each of the triangular mel weighting filters.

$$s(m) = \sum_{k=0}^{N-1}\left[|X(k)|^2 H_m(k)\right]; \qquad 0 \le m \le M - 1 \qquad (A.4)$$

where M is the total number of triangular mel weighting filters [128,129]. $H_m(k)$ is the weight given to the kth energy spectrum bin contributing to the mth output band and is expressed as:

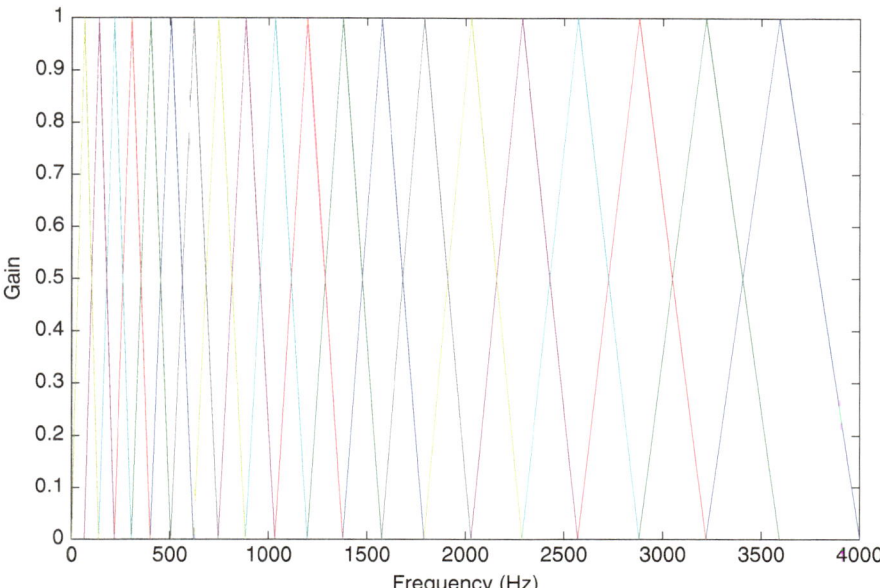

Fig. A.1 Mel-filter bank

$$
H_m(k) = \begin{cases}
0, & k < f(m-1) \\
\frac{2(k-f(m-1))}{f(m)-f(m-1)}, & f(m-1) \le k \le f(m) \\
\frac{2(f(m+1)-k)}{f(m+1)-f(m)}, & f(m) < k \le f(m+1) \\
0, & k > f(m+1)
\end{cases} \tag{A.5}
$$

with m ranging from 0 to $M-1$.

5. *Discrete Cosine Transform (DCT):* Since the vocal tract is smooth, the energy levels in adjacent bands tend to be correlated. DCT applied to the transformed mel frequency coefficients produces a set of cepstral coefficients. Prior to computing DCT the mel spectrum is usually represented on a log scale. This results in a signal in the cepstral domain with a que-frequency peak corresponding to the pitch of the signal and a number of formants representing low que-frequency peaks. Since most of the signal information is represented by the first few MFCC coefficients, the system can be made robust by extracting only those coefficients ignoring or truncating higher order DCT components [101]. Finally, MFCC is calculated as [101]

$$
c(n) = \sum_{m=0}^{M-1} \log_{10}(s(m)) \cos\left(\frac{\pi n(m-0.5)}{M}\right); \qquad n = 0, 1, 2, \ldots, C-1 \tag{A.6}
$$

where $c(n)$ are the cepstral coefficients and C is the number of MFCCs. Traditional MFCC systems use only 8–13 cepstral coefficients. The zeroth coefficient is often excluded since it represents the average log-energy of the input signal, which only carries little speaker-specific information.

6. *Dynamic MFCC features:* The cepstral coefficients are usually referred to as static features, since they only contain information from a given frame. The extra information about the temporal dynamics of the signal is obtained by computing first and second derivatives of cepstral coefficients [130–132]. The first order derivative is called delta coefficients, and the second order derivative is called delta–delta coefficients. Delta coefficients tell about the speech rate, and delta–delta coefficients provide information similar to acceleration of speech. The commonly used definition for computing dynamic parameter is [130]

$$\Delta c_m(n) = \frac{\sum\limits_{i=-T}^{T} k_i c_m(n+i)}{\sum\limits_{i=-T}^{T} |i|} \tag{A.7}$$

where $c_m(n)$ denotes the mth feature for the nth time frame, k_i is the ith weight, and T is the number of successive frames used for computation. Generally T is taken as 2. The delta–delta coefficients are computed by taking the first order derivative of the delta coefficients.

Appendix B
Speech Coders

B.1 Global System for Mobile Full Rate Coder (ETSI GSM 06.10)

GSM full rate coder provides 13 kbps bit rate using regular pulse excitation and long-term prediction (RPE-LTP) techniques. LTP captures the long-term correlations present in the speech signal [133]. GSM full rate speech encoder takes its input as a 13-bit uniform pulse code modulation (PCM) signal sampled at 8 kHz. The input PCM signal is processed on a frame-by-frame basis, with a frame size of 20 ms (160 samples). Bits allocation of GSM full rate coder is shown in Table B.1. Full rate GSM was the first digital speech coding standard used in the GSM digital mobile phone system.

B.2 GSM Enhanced Full Rate Coder (ETSI GSM 06.60)

GSM enhanced full rate (EFR) coder is developed to improve the performance of GSM full rate coder under severe error conditions. GSM EFR coder provides 12.2 kbps bit rate by using algebraic code excited linear prediction (ACELP) scheme. It operates on 20 ms frames of speech, sampled at 8 kHz. GSM EFR coder is compatible with highest adaptive multi-rate (AMR) coder. Bits allocation of GSM EFR coder is shown in Table B.2.

B.3 Codebook Excited Linear Prediction (CELP FS-1016)

CELP is based on the concept of linear predictive coding (LPC). LPC estimates the current speech sample by the linear combination of the past speech samples. In CELP, a codebook of different excitation signals is maintained at the encoder and

K.S. Rao and A.K. Vuppala, *Speech Processing in Mobile Environments*, SpringerBriefs in Electrical and Computer Engineering, DOI 10.1007/978-3-319-03116-3,
© Springer International Publishing Switzerland 2014

Table B.1 GSM FR (ETSI
GSM 06.10) bits allocation

Parameter to be encoded	Number of bits
LPC	36
Pitch period	28
Long term gain D	8
Position	8
Peak magnitude	24
Sample amplitude	156
Total no. of bits per frame	260
Bit rate	13 kbps

Table B.2 GSM EFR (ETSI
GSM 06.60) bits allocation

Parameter to be encoded	Number of bits
LPC	38
Pitch period	30
Adaptive code book gain	16
Algebraic code book index	140
Algebraic code book gain	20
Total no. of bits per frame	244
Bit rate	12.2 kbps

Table B.3 CELP (FS-1016)
bits allocation

Parameter to be encoded	Number of bits
LPC coefficients (10 LSP's)	34
Pitch prediction	48
Code book	36
Gains	20
Synchronization	1
FEC	4
Frame expansion	1
Total no. of bits per frame	144
Bit rate	4.8 kbps

decoder. The encoder finds the most suitable excitation signal and sends its index to the decoder, which then uses it to reproduce the signal [133, 134]. Hence the name codebook excited is suggested to this coder. CELP FS-1016 operates at a bit rate of 4.8 kbps. CELP is a widely used speech coding algorithm, and one of the practical application of it is in selective mode vocoder (SMV) for CDMA. Bits allocation for CELP FS-1016 coder is shown in Table B.3.

B.4   Mixed Excited Linear Prediction (MELP TI 2.4 kbps)

MELP utilizes more sophisticated speech production model, with additional parameters to capture the underlying signal dynamics with improved accuracy. Excitation signal is generated by combining the filtered periodic pulse sequence with the filtered noise sequence [133]. MELP (TI 2.4 kbps) operates at a bit rate of 2.4 kbps.

Table B.4 MELP (TI
2.4 kbps) bits allocation

Parameter to be encoded	Number of bits
LPC coefficients (10 LSP's)	34
Gain (2 per frame)	8
Pitch and overall voicing	7
Band-pass voicing	5-1
Aperiodic flag	1
Total no. of bits per frame	54
Bit rate	2.4 kbps

Table B.5 Comparison
of speech coders

Algorithm	Bit-rate (kbps)	MOS	Complexity (MIPS)	Frame size (ms)
PCM	64	4.3	0.01	0
GSM FR	13	3.5–3.9	5–6	20
GSM EFR	12.2	3.8	14	20
CELP	4.8	3.2	16	30
MELP	2.4	3.2	40	22.5

MELP is used in military, satellite, and secure voice applications. Allocation of bits
for MELP coder is shown in Table B.4.

Comparison of speech coders considered in this study in terms of complexity
in million instructions per second (MIPS), mean opinion score (MOS), and input
frame size is shown in Table B.5. We can observe in Table B.5 that as coding rate
decreases complexity increases.

B.5   Degradation Measures

Degradation introduced by speech coding is measured using log-likelihood ratio,
weighted spectral slope measure, and log-spectral distance. They are defined as
follows.

- Log-likelihood ratio (LLR) [135]

$$LLR = log_{10} \left[\frac{a_x R_x a_x^T}{a_y R_y a_y^T} \right] \tag{B.1}$$

where a_x and a_y are LP coefficient vectors of the original and coded (degraded)
speech. a_x^T and a_y^T are transpose of LP coefficient vectors of the original and
coded (degraded) speech. R_x and R_y are autocorrelation matrices of the original
and coded (degraded) speech.
- Weighted spectral slope (WSS) [135]

Table B.6 Comparison of quality measures for different speech coders

Algorithm	LLR	WSS	LSD
GSM FR (ETSI 06.10)	0.17	12.8	0.88
GSM EFR (ETSI 06.60)	0.16	11.9	0.84
CELP (FS1016)	0.54	51.2	1.27
MELP (TI 2.4 kbps)	0.33	37	1.12

The WSS measure is based on auditory model in which 36 overlapping filters of progressively larger bandwidth are used to estimate the smoothed short-time speech spectrum

$$WSS = K_{spl}(k - \hat{k}) + \sum_{k=1}^{36} W_a(k)[S(k) - \hat{S}(k)]^2 \qquad (B.2)$$

where k, $\hat{k}$ are related to overall sound pressure level of the original and coded (degraded) utterances. K_{spl} is a parameter which can be varied to increase overall performance. $W_a(k)$ is the weight of each band. $S(k)$ and $\hat{S}(k)$ are the slopes in each critical band k for the original and coded (degraded) speech utterances.

• Log-spectral distance (LSD) [135]

$$LSD = \sqrt{\frac{1}{2\pi} \int_{-\pi}^{\pi} \left[10 \log_{10} \frac{P(\omega)}{\hat{P}(\omega)} \right]^2 d\omega}. \qquad (B.3)$$

where $P(\omega)$ and $\hat{P}(\omega)$ are power spectrum of clean and coded (degraded) speech.

Above quality measures are calculated for different speech utterances from TIMIT database, and average measures are shown in Table B.6. In Table B.6 column-1 indicates different coders considered in this study; Columns 2–4 indicate the average LLR, WSS, and LSD values, respectively, for different coders. Results indicate that CELP coder introduces more degradation among the coders considered.

Appendix C
Pattern Recognition Models

In this work hidden Markov model (HMM), support vector machine (SVM), and auto-associative neural network (AANN) models are used to capture the pattern present in features. HMMs are used to capture the sequential information present in feature vectors for CV recognition. SVMs are used to capture the discriminative information present in the feature vectors for CV recognition. AANN models are used to capture the nonlinear relations among the feature vectors for speaker identification. The following sections briefly describe the pattern recognition models used in this study.

C.1 Hidden Markov Models

Hidden Markov models (HMMs) are the commonly used classification models in speech recognition [136]. HMM is a stochastic signal model which is referred to as Markov sources or probabilistic functions of Markov chains. This model is an extension to the concept of Markov model which includes the case where the observation is a probabilistic function of the state. HMM is a finite set of states, each of which is associated with a probability distribution. Transitions among the states are governed by a set of probabilities called transition probabilities. In a particular state an outcome or observation can be generated, according to the associated probability distribution. It is only the outcome is known and underlying state sequence is hidden. Hence, it is called a hidden Markov model.

Following are the basic elements that define HMM:

1. N, The number of states in the model,
   $s = \{s_1, s_2, \ldots \ldots s_N\}$
2. M, Number of distinct observation symbol per state,
   $v = \{v_1, v_2, \ldots . v_M\}$

3. State transition probability distribution A = $\{a_{ij}\}$ where

$$a_{ij} = P\left[q_{t+1} = s_j \,|\, q_t = s_i\right], 1 \le i, j \le N \qquad (C.1)$$

4. Observation symbol probability distribution in state j,
   B = $\{ b_j(\text{k}) \}$ where

$$b_j(k) = P\left[v_k \text{ at } t \,|\, q_t = s_j\right] \quad 1 \le j \le N, 1 \le k \le M \qquad (C.2)$$

5. Initial state distribution $\Pi = \{\Pi_j\}$ where

$$\Pi_j = P\left[q_1 = s_i\right] \quad 1 \le i \le N \qquad (C.3)$$

So, a complete specification of an HMM requires specification of two model parameters (N and M) , specification of observation symbols, and the specification of three probability measures A,B,Π. Therefore HMM is indicated by the compact notation

$$\lambda = (A, B, \Pi)$$

Given that state sequence $q = (q1q2\ldots q_T)$ is unknown, the probability of observation sequence $O = (o1o2\ldots o_T)$ given the model λ is obtained by summing the probability of over all possible state sequences q as follows:

$$P(o|\lambda) = \sum_{q_1, q_2, \ldots, q_T} \pi_{q_1} b_{q_1}(o_1) a_{q_1 q_2} b_{q_2}(o_2) \ldots s a_{q_{T-1} q_T} b_{q_T}(o_T) \qquad (C.4)$$

where π_{q_1} is the initial state probability of q_1 and T is the length of observation sequence.

C.2   Support Vector Machines

A notable characteristic of a support vector machine (SVM) is that the computational complexity is independent of the dimensionality of the kernel space, where the input feature space is mapped. Thus, the curse-of-dimensionality is bypassed in SVM. SVMs have been applied to number of different applications ranging from handwritten digit recognition to person identification. The results shown in these studies indicate that SVM classifiers exhibit enhanced generalization performance [137]. However, intelligent design or choice of kernel function adds to the real strength of support vector machines. SVMs are designed for two-class pattern classification. Multiclass (n-class) pattern classification problems can be solved using a combination of binary (2-class) support vector machines. One-against-the-rest approach is used for decomposition of n-class pattern classification problem into n two-class classification problems. The set of training examples $\left\{\{(x_i, k)\}_{i=1}^{N_k}\right\}_{k=1}^{n}$

Fig. C.1 Classification mechanism in support vector machines

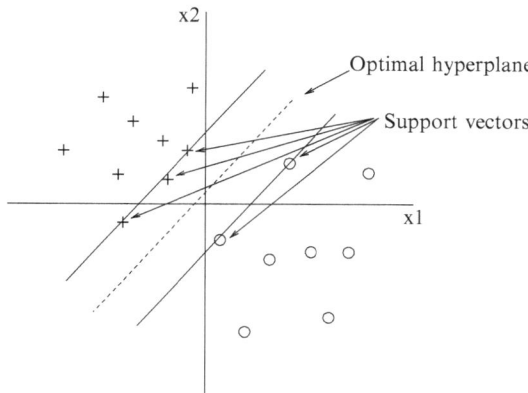

consists of N_k number of examples belonging to the kth class, where the class label $k \in \{1, 2, 3, \ldots, n\}$. All the training examples are used to construct the SVM for a class. The SVM for the class k is constructed using the set of training examples and their desired outputs, $\left\{ \{(x_i, y_i)\}_{i=1}^{N_k} \right\}_{k=1}^{n}$. The desired output y_i for the training example x_i is defined as follows:

$$y_i = \begin{cases} +1 \text{ if } x_i \in kth \text{ class} \\ -1 \text{ otherwise} \end{cases}$$

The examples with $y_i = +1$ are called positive examples and those with $y_i = -1$ are negative ones. An optimal hyperplane is constructed to separate positive examples from negative ones. The separating hyperplane (margin) is chosen in such a way as to maximize its distance from the closest training examples of different classes. Figure C.1 illustrates the geometric construction of hyperplane for two-dimensional input space. The support vectors are those data points that lie closest to the decision surface and therefore the most difficult to classify. They have a direct bearing on the optimum location of the decision surface. For a given test pattern x, the evidence $D_k(x)$ is obtained from each of the SVMs. In the decision logic, the class label k associated with SVM, which gives maximum evidence, is hypothesized as the class (C) of the test pattern, that is

$$C(x) = argmax(D_k(x))$$

C.3 Auto-Associative Neural Network Models

AANN models are basically feed-forward neural network (FFNN) models, which try to map an input vector onto itself, and hence the name auto-association or identity mapping [117, 118]. It consists of an input layer, an output layer, and one or more hidden layers. The number of units in the input and output layers is equal to the dimension of the input feature vectors. The number of nodes in one of the hidden layers is less than the number of units in either the input or the output layer. This hidden layer is also known as dimension compression layer. The activation function of the units in the input and output layers is linear, whereas in case of hidden layers it is either linear or nonlinear.

A five-layer AANN model with the structure shown in Fig. 6.1 is used in this study. The structure of network used to capture the higher order relations is 39L 60N 20N 60N 39L, where L refers to linear units and N to nonlinear units. The integer value indicates the number of units present in that layer. Number of linear elements at the input layer indicates size of the feature vectors used for developing the models. The nonlinear units use $tanh(s)$ as the activation function, where s is the activation value of that unit. The structure of the network was determined empirically. The back-propagation learning algorithm is used for adjusting the weights of the network to minimize the mean squared error. The performance of AANN models can be interpreted as linear or nonlinear principal component analysis (PCA) or capturing the distribution of input data [117, 118].

Determining the network structure is an optimization problem. At present there are no formal methods for determining the optimal structure of a neural network. The key factors that influence the neural network structure are learning ability of a network and capacity to generalize the acquired knowledge. From the available literature, it is observed that five-layer symmetric neural networks with three hidden layers have been used for different speech tasks. The first and the third hidden layers have more number of nodes than the input or output layer. Middle layer (also known as dimension compression layer) contains less number of units. In this type of network, generally the first and third hidden layers are expected to capture the local information among the feature vectors and the middle hidden layer is meant for capturing global information. The five-layer AANNs structure for their optimal performance is $N_1 L - N_2 N - N_3 N - N_2 N - N_1 L$. Here N_1, N_2, and N_3 indicate the number of units in the first, second, and third layers, respectively, of the symmetric five-layer AANN. Usually N_2 and N_3 are derived experimentally for achieving the best performance in the given task. From the existing studies, it is observed that N_2 is in the range of 1.3–2 times of N_1 and N_3 is in the range of 0.2–0.6 times of N_1.

References

1. K.N. Stevens, *Acoustic Phonetics* (MIT Press, Cambridge, MA, 1999)
2. D. Crystal, *A Dictionary of Linguistics and Phonetics* (Basil Blackwell, Cambridge, Massachusetts, 1985)
3. M.A. Jack, J. Laver, *Aspects of Speech Technology* (Edinburgh university press, Edinburgh, 1988)
4. S.R.M. Prasanna, Event-based analysis of speech, PhD thesis, IIT Madras, March 2004
5. S.R.M. Prasanna, S.V. Gangashetty, B. Yegnanarayana, Significance of vowel onset point for speech analysis, in *Proc. of Int. Conf. Signal Processing and Communications*, (Bangalore, India, 2001), pp. 81–88
6. K.S. Rao, Voice conversion by mapping the speaker-specific features using pitch synchronous approach. Comput. Speech Lang. **24**, 474–494 (2010)
7. D.J. Hermes, Vowel onset detection. J. Acoust. Soc. Am. **87**, 866–873 (1990)
8. J.-H. Wang, S.-H. Chen, A C/V segmentation algorithm for Mandarin speech using wavelet transforms, in *Proc. IEEE Int. Conf. Acoust., Speech, Signal Processing* (Phoenix, Arizona, 1999), pp. 1261–1264
9. S.V. Gangashetty, C.C. Sekhar, B. Yegnanarayana, Detection of vowel onset points in continuous speech using autoassociative neural network models, in *Proc. Int. Conf. Spoken Language Processing*, (Jeju Island, Korea, 2004), pp. 401–410
10. J.-F. Wang, C.H. Wu, S.H. Chang, J.Y. Lee, A hierarchical neural network based C/V segmentation algorithm for Mandarin speech recognition. IEEE Trans. Signal Process. **39**(9), 2141–2146 (1991)
11. S.V. Gangashetty, C.C. Sekhar, B. Yegnanarayana., Extraction of fixed dimension patterns from varying duration segments of consonant-vowel utterances, in *Proc. of IEEE ICISIP*, pp. 159–164, 2004
12. S.R.M. Prasanna, B. Yegnanarayana, Detection of vowel onset point events using excitation source information, in *Proc. of Interspeech* (Lisbon, Portugal, 2005), pp. 1133–1136
13. A. Kazemzadeh, J. Tepperman, J. Silva, H. You, S. Lee, A. Alwan, S. Narayanan, Automatic detection of voice onset time contrasts for use in pronunciation assessment, in *Proc. Int. Conf. Spoken Language Processing* (Pittsburgh, PA, USA, 2006)
14. V. Stouten, H.V. hamme, Automatic voice onset time estimation from reassignment spectra. Speech Comm. **51**, 1194–1205 (2009)
15. S.R.M. Prasanna, B.V.S. Reddy, P. Krishnamoorthy, Vowel onset point detection using source, spectral peaks, and modulation spectrum energies. IEEE Trans. Audio Speech Lang. Process. **17**, 556–565 (2009)
16. K.S. Rao, B. Yegnanarayana, Duration modification using glottal closure instants and vowel onset points. Speech Comm. **51**, 1263–1269 (2009)

K.S. Rao and A.K. Vuppala, *Speech Processing in Mobile Environments*, SpringerBriefs   115
in Electrical and Computer Engineering, DOI 10.1007/978-3-319-03116-3,
© Springer International Publishing Switzerland 2014

17. K.S. Rao, A.K. Vuppala, Non-uniform time scale modification using instants of significant excitation and vowel onset points. Speech Comm. (Elsevier) **55**(6), 745–756 (2013)
18. J.H.L. Hansen, S.S. Gray, W. Kim, Automatic voice onset time detection for unvoiced stops (/p/,/t/,/k/) with application to accent classification. Speech Comm. **52**, 777–789 (2010)
19. C. Prakash, N. Dhananjaya, S. Gangashetty, Bessel features for detection of voice onset time using AM-FM signal, in *Proc. of Int. Conf. on the Systems, Signals and Image Processing (IWSSIP)*, (IEEE, Sarajevo, Bosnia and Herzegovina, 2011), pp. 1–4
20. D. Zaykovskiy, Survey of the speech recognition techniques for mobile devices, in *Proc. of DS Publications*, 2006
21. Z.H. Tan, B. Lindberg, *Automatic Speech Recognition on Mobile Devices and over Communication Networks* (Springer, London, 2008)
22. J.M. Huerta, Speech recognition in mobile environments, PhD thesis, Carnegie Mellon University, Apr. 2000
23. A.M. Peinado, J.C. Segura, *Speech Recognition over Digital Channels* (Wiley, New York, 2006)
24. S. Kafley, A.K. Vuppala, A. Chauhan, K.S. Rao, "Continuous digit recognition in mobile environment," in *Proc. of IEEE Techsym* (IIT Kharagpur, India, 2010), pp. 217–222
25. A.M. Gomez, A.M. Peinado, V. Sanchez, A.J. Rubio, Recognition of coded speech transmitted over wireless channels. IEEE Trans. Wireless Comm. **5**, 2555–2562 (2006)
26. S. Euler, J. Zinke, The influence of speech coding algorithms on automatic speech recognition, in *Proc. IEEE Int. Conf. Acoust., Speech, Signal Processing* (Adelaide, Australia, 1994), pp. 621–624
27. B.T. Lilly, K.K. Paliwal, Effect of speech coders on speech recognition performance, in *Proc. Int. Conf. Spoken Language Processing* (Philadelphia, PA, USA, 1996), pp. 2344–2347
28. A. Gallardo-Antolin, C. Pelaez-Moreno, F.D. de Maria, Recognizing GSM digital speech. IEEE Trans. Speech Audio Process **13**(6), 1186–1205 (2005)
29. F. Quatieri, E. Singer, R.B. Dunn, D.A. Reynolds, J.P. Campbell, Speaker and language recognition using speech codec parameters, in *Proc. of Eurospeech* (Budapest, Hungary, 1999), pp. 787–790
30. R.B. Dunn, T.F. Quatieri, D.A. Reynolds, J.P. Campbell, Speaker recognition from coded speech in matched and mismatched condition, in *Proc. of Speaker Recognition Workshop* (Crete, Greece, 1999), pp. 115–120
31. R. Dunn, T. Quatieri, D. Reynolds, J. Campbell, Speaker recognition from coded speech and the effects of score normalization, in *Proc. of Thirty-Fifth Asilomar Conference on Signals, Systems and Computers* (IEEE, Monterery, CA, USA, 2001), pp. 1562–1567
32. A. Krobba, M. Debyeche, A. Amrouche, Evaluation of speaker identification system using GSM-EFR speech data, in *Proc. of Int. Conf. on Design and Technology of Integrated Systems* (Nanoscale Era Hammamet, 2010), pp. 1–5
33. A. Janicki, T. Staroszczyk, Speaker recognition from coded speech using support vector machines, in *Proc. of 4th Int. Conf. on Text, Speech and Dialogue* (Springer, Pilsen, Czech Republic, 2011), pp. 291–298
34. C. Mokbel, G. Chollet, Speech recognition in adverse environments: speech enhancement and spectral transformations, in *Proc. IEEE Int. Conf. Acoust., Speech, Signal Processing* (Toronto, Ontario, Canada, 1991)
35. J.A. Nolazco-Flores, S. Young, *CSS-PMC: a combined enhancement/compensation scheme for continuous speech recognition in noise*. Cambridge University Engineering Department. Technical Report, 1993
36. J. Huang, Y. Zhao, Energy-constrained signal subspace method for speech enhancement and recognition. IEEE Signal Process. Lett. **4**, 283–285 (1997)
37. K. Hermus, W. Verhelst, P. Wambacq, Optimized subspace weighting for robust speech recognition in additive noise environments, in *Proc. of ICSLP* (Beijing, China, 2000), pp. 542–545

38. K. Hermus, P. Wambacq, Assessment of signal subspace based speech enhancement for noise robust speech recognition, in *Proc. IEEE Int. Conf. Acoust., Speech, Signal Processing* (Montreal, Canada, 2004), pp. 945–948

39. H. Kris, W. Patrick, V.H. Hugo, A review of signal subspace speech enhancement and its application to noise robust speech recognition. EURASIP J. Appl. Signal Process. 195–209 (2007)

40. H. Hermanski, N. Morgan, H.G. Hirsch, Recognition of speech in additive and convolutional noise based on RASTA spectral processing, in *Proc. IEEE Int. Conf. Acoust., Speech, Signal Processing* (Adelaide, Australia, 1994)

41. O. Viiki, B. Bye, K. Laurila, A recursive feature vector normalization approach for robust speech recognition in noise, in *Proc. IEEE Int. Conf. Acoust., Speech, Signal Processing* (Seattle, USA, 1998)

42. D. Yu, L. Deng, J. Droppo, J. Wu, Y. Gong, A. Acero, A minimum-mean-square-error noise reduction algorithm on mel-frequency cepstra for robust speech recognition, in *Proc. IEEE Int. Conf. Acoust., Speech, Signal Processing*, (Las Vegas, USA, 2008), pp. 4041–4044

43. X. Cui, A. Alwan, Noise robust speech recognition using feature compensation based on polynomial regression of utterance SNR. IEEE Trans. Speech Audio Process. **13**, 1161–1172 (2005)

44. F. Hilger, H. Ney, Quantile based histogram equalization for noise robust large vocabulary speech recognition. IEEE Trans. Audio Speech Lang. Process. **14**(3), 845–854 (2006)

45. A. de la Torre, A.M. Peinado, J.C. Segura, J.L. Perez-Cordoba, M.C. Benitez, A.J. Rubio, Histogram equalization of speech representation for robust speech recognition. IEEE Trans. Speech Audio Process. **13**(3), 355–366 (2005)

46. Y. Suh, M. Ji, H. Kim, Probabilistic class histogram equalization for robust speech recognition. IEEE Signal Process. Lett. **14**(4), 287–290 (2007)

47. K. Ohkura, M. Sugiyama, Speech recognition in a noisy environment using a noise reduction neural network and a codebook mapping technique, in *Proc. IEEE Int. Conf. Acoust., Speech, Signal Processing* (Toronto, Canada, 1991)

48. M. Gales, S.Young, S.J. Young, Robust continuous speech recognition using parallel model combination. IEEE Trans. Speech Audio Process. **4**(5), 352–359 (1996)

49. P.J. Moreno, *Speech Recognition in Noisy Environments*, PhD thesis, Carnegie Mellon University, 1996

50. S.V. Vaseghi, B.P. Milner, Noise compensation methods for hidden Markov model speech recognition in adverse environments. IEEE Trans. Speech Audio Process. **5**, 11–21 (1997)

51. H. Liao, M.J.F. Gales, Adaptive training with joint uncertainty decoding for robust recognition of noisy data, in *Proc. IEEE Int. Conf. Acoust., Speech, Signal Processing* (Honolulu, USA, 2007), pp. 389–392

52. O. Kalinli, M.L. Seltzer, J. Droppo, A. Acero, Noise adaptive training for robust automatic speech recognition. IEEE Trans. Audio, Speech Lang. Process. **18**(8), 1889–1901 (2010)

53. D.K. Kim, M.J.F. Gales, Noisy constrained maximum-likelihood linear regression for noise-robust speech recognition. IEEE Trans. Audio Speech Lang. Process. **19**(2), 315–325 (2011)

54. S.V. Gangashetty, *Neural network models for recognition of consonant-vowel units of speech in Multiple Languages*, PhD thesis, IIT Madras, October 2004

55. C.C. Sekhar, *Neural Network models for recognition of stop consonant-vowel (SCV) segments in continuous speech*, PhD thesis, IIT Madras, 1996

56. K.S. Rao, Application of prosody models for developing speech systems in indian languages. Int. J. Speech Tech. (Springer) **14**, 19–33 (2011)

57. C.C. Sekhar, W.F. Lee, K. Takeda, F. Itakura, Acoustic modeling of subword units using support vector machines, in *Proc. of WSLP* (Mumbai, India, 2003)

58. S.V. Gangashetty, C.C. Sekhar, B. Yegnanarayana, Combining evidence from multiple classifiers for recognition of consonant-vowel units of speech in multiple languages, in *Proc. of ICISIP* (Chennai, India, 2005), pp. 387–391

59. K.S. Rao, B. Yegnanarayana, Prosody modification using instants of significant excitation. IEEE Trans. Audio Speech Lang. Process. **14**, 972–980 (2006)

60. E. Moulines, J. Laroche, Non-parametric techniques for pitch-scale and time-scale modification of speech. Speech Comm. **16**, 175–205 (1995)
61. M.R. Portnoff, Time-scale modification of speech based on short-time Fourier analysis. IEEE Trans. Acoust. Speech Signal Process. **29**, 374–390 (1981)
62. H.G. Ilk, S. Guler, Adaptive time scale modification of speech for graceful degrading voice quality in congested networks for VoIP applications. Signal Process. **86**, 127–139 (2006)
63. K.S. Rao, Real time prosody modification. J. Signal Inform. Process. 50–62 (2010)
64. T.F. Quatieri, R.J. McAulay, Shape invariant time-scale and pitch modification of speech. IEEE Signal Process. **40**, 497–510 (1992)
65. J. di Marino, Y. Laprie, Supression of phasiness for time-scale modifications of speech signals based on a shape invarience property, in *Proc. IEEE Int. Conf. Acoust., Speech, Signal Processing* (Saltlake city, Utah, USA, 2001)
66. E. Moulines, F. Charpentier, Pitch-synchronous waveform processing techniques for text to speech synthesis using diphones. Speech Comm. **9**, 453–467 (1990)
67. M. Slaney, M. Covell, B. Lassiter, Automatic audio morphing, in *Proc. IEEE Int. Conf. Acoust., Speech, Signal Processing* (Atlanta, GA, USA, 1996)
68. O. Donnellan, E. Jung, E. Coyle, Speech-adaptive time-scale modification for computer assisted language-learning, in *Proc. of 3rd IEEE Int. Conf. on Advanced Learning Technologies (ICALT03)* (Aix-en-Provence, France, 2003)
69. A. Klapuri, Sound onset detection by applying psychoacoustic knowledge, in *Proc. IEEE Int. Conf. Acoust., Speech, Signal Processing* (Washington, DC, USA, 1999), pp. 3089–3092
70. C. Duxbury, M.E. Davies, M.B. Sandler, Separation of transient information in musical audio using multiresolution analysis techniques, in *Proc. of Int. Conf. Digital Audio Effects (DAFX) Limerick* (Limerick, 2001), pp. 1–4
71. J. Bonada, Automatic technique in frequency domain for near-lossless time-scale modification of audio, in *Proc. of Int. Conf. Computer Music Conference (ICMC)* (Berlin, Germany, 2000), pp. 396–399
72. C. Duxbury, M.E. Davies, M. Sandler, Improved time-scaling of musical audio using phase locking at transients, in *Proc. of Audio Engineering Society Convention 11* (Munich, Germany, 2002), paper 5530
73. A. Roebel, A new approach to transient processing in the phase vocoder, in *Proc. of Int. Conf. Digital Audio Effects (DAFX)* (London, 2003), pp. 344–349
74. X. Rodet, F. Jaillet, Detection and modeling of fast attack transients, in *Proc. of Int. Conf. Computer Music Conference (ICMC)* (Havana, Cuba, 2001), pp. 30–33
75. S. Hainsworth, M. Macleod, P. Wolfe, Analysis of reassigned spectrograms for musical transcription, in *Proc. of IEEE Workshop on Applications of Signal Processing to Audio and Acoustics* (New Paltz, NY, 2001), pp. 23–26
76. S. Grofit, Y. Lavner, Time-scale modification of audio signals using enhanced WSOLA with management of transients. IEEE Trans. Audio Speech Lang. Process. **16**, 106–115 (2008)
77. J.S. Garofolo, L.F. Lamel, W.M. Fisher, J.G. Fiscus, D.S. Pallett, N.L. Dahlgren, V. Zue, TIMIT acoustic-phonetic continuous speech corpus linguistic data consortium, in *Proc. of IEEE ICISIP* (Philadelphia, PA, 1993)
78. S.V. Gangashetty, C.C. Sekhar, B. Yegnanarayana, Spotting multilingual consonant-vowel units of speech using neural networks, in *An ISCA Tutorial and Research Workshop on Nonlinear Speech Processing*, pp. 287–297, 2005
79. R.M. Hegde, H.A. Murthy, V. Gadde, Continuous speech recognition using joint features derived from the modified group delay function and MFCC, in *Proc. of INTERSPEECH-Int. Conf. Spoken Language Processing* (Jeju Island, Korea, 2004), pp. 905–908
80. K.S. Rao, B. Yegnanarayana, Intonation modeling for Indian languages. Comput. Speech Lang. **23**, 240–256 (2009)
81. K.S. Rao, B. Yegnanarayana, Modeling durations of syllables using neural networks. Comput. Speech Lang. (Elsevier) **21**, 282–295 (2007)
82. K.S. Rao, S.G. Koolagudi, Selection of suitable features for modeling the durations of syllables. J. Softw. Eng. Appl. 1107–1117 (2010)

83. K.S. Rao, Role of neural network models for developing speech systems. SADHANA (Springer) **36**, 783–836 (2011)
84. L. Mary, K.S. Rao, B. Yegnanarayana, Neural Network Classifiers for Language Identification using Syntactic and Prosodic features, in *Proc. IEEE Int. Conf. Intelligent Sensing and Information Processing* (Chennai, India, 2005), pp. 404–408
85. L. Mary, B. Yegnanarayana, Extraction and representation of prosodic features for language and speaker recognition. Speech Comm. **50**, 782–796 (2008)
86. K.S. Rao, *Acquisition and incorporation of prosody knowledge for speech systems in indian languages*, PhD thesis, Department of Computer Science and Engineering, Indian Institute of Technology Madras, May 2005
87. A.K. Vuppala, J. Yadav, K.S. Rao, S. Chakrabarti, Vowel onset point detection for low bit rate coded speech. IEEE Trans. Audio Speech Lang. Process. **20**(6), 1894–1903 (2012)
88. S.R.M. Kodukula, *Significance of excitation source information for speech analysis*. PhD thesis, IIT Madras, March 2009
89. S. Guruprasad, Exploring features and scoring methods for speaker recognition, Master's thesis, MS Thesis, IIT Madras, 2004
90. P.S. Murthy, B. Yegnanarayana, Robustness of group-delay-based method for extraction of significant instants of excitation from speech signals. IEEE Trans. Speech Audio Process. **7**, 609–619 (1999)
91. K.S. Rao, S.R.M. Prasanna, B. Yegnanarayana, Determination of instants of significant excitation in speech using hilbert envelope and group delay function. IEEE Signal Process. Lett. **14**, 762–765 (2007)
92. K.S.R. Murty, B. Yegnanarayana, Epoch extraction from speech signals. IEEE Trans. Audio Speech Lang. Process. **16**(8), 1602–1613 (2008)
93. A.K. Vuppala, J. Yadav, K.S. Rao, S. Chakrabarti, Effect of speech coding on epoch extraction, in *Proc. of IEEE Int. Conf. on Devices and Communications*, (Mesra, India, 2011)
94. A.K. Vuppala, K.S. Rao, S. Chakrabarti, Vowel onset point detection for noisy speech using spectral energy at formant frequencies. Int. J. Speech Tech. (Springer) **16**(2), 229–235 (2013)
95. M.A. Joseph, S. Guruprasad, B. Yegnanarayana, Extracting formants from short segments of speech using group delay functions, in *Proc. of Interspeech* (Pittsburgh, PA, USA, 2006), pp. 1009–1012
96. M.A. Joseph, Extracting formant frequencies from short segments of speech, Master's thesis, Dept. of Computer Science and Engineering, Indian Institute of Technology Madras, Apr. 2008
97. Noisex-92: http://spib.rice.edu/spib/select_noise.html
98. A.K. Vuppala, J. Yadav, K.S. Rao, S. Chakrabarti, Effect of noise on vowel onset point detection, in *Proc. of Int. Conf. Contemporary Computing* (Noida, India, 2011), pp. 201–211. Communications in Computer and Information Science (Springer)
99. A.K. Vuppala, S. Chakrabarti, K.S. Rao, Effect of speech coding on recognition of consonant-vowel (CV) units, in *Proc. of Int. Conf. contemporary computing (Springer Communications in Computer and Information Science ISSN: 1865–0929)*, (Noida, India, 2010), pp. 284–294
100. A.K. Vuppala, K.S. Rao, S. Chakrabarti, Improved consonant-vowel recognition for low bit-rate coded speech. Wiley Int. J. Adapt. Contr. Signal Process. **26**, 333–349 (2012)
101. J.W. Picone, Signal modeling techniques in speech recognition. Proc. IEEE **81**, 1215–1247 (1993)
102. S. Young, D. Kershaw, J. Odell, D. Ollason, V. Valtchev, P. Woodland, *The HTK Book Version 3.0* (Cambridge University Press, Cambridge, 2000)
103. R. Collobert, S. Bengio, SVMTorch: support vector machines for large-scale regression problems. Proc. J. Mach. Learn. Res. 143–160 (2001)
104. A.K. Vuppala, K.S. Rao, S. Chakrabarti, Improved vowel onset point detection using epoch intervals. AEUE (Elsevier) **66**, 697–700 (2012)
105. P. Krishnamoorthy, S.R.M. Prasanna, Enhancement of noisy speech by temporal and spectral processing. Speech Comm. **53**, 154–174 (2011)

106. S. Bell, Suppression of acoustic noise in speech using spectral subtraction. IEEE Trans. Acoust. Speech Signal Process. **27**, 113–120 (1979)

107. S. Kamath, P. Loizou, A multi-band spectral subtraction method for enhancing speech corrupted by colored noise, in *Proc. IEEE Int. Conf. Acoust., Speech, Signal Processing* (Orlando, USA, 2002)

108. Y. Ephrain, D. Malah, Speech enhancement using minimum mean square error short-time spectral amplitude estimator. IEEE Trans. Acoust. Speech Signal Process. **32**, 1109–1121 (1984)

109. B. Yegnanarayana, C. Avendano, H. Hermansky, P.S. Murthy, Speech enhancement using linear prediction residual. Speech Comm. **28**, 25–42 (1999)

110. B. Yegnanarayana, P.S. Murthy, Enhancement of reverberant speech using lp residual signal. IEEE Trans. Speech Audio Process. **8**, 267–281 (2000)

111. B. Yegnanarayana, S.R.M. Prasanna, R. Duraiswami, D. Zotkin, Processing of reverberant speech for time-delay estimation. IEEE Trans. Speech Audio Process. **13**, 1110–1118 (2005)

112. A.K. Vuppala, K.S. Rao, S. Chakrabarti, P. Krishnamoorthy, S.R.M. Prasanna, Recognition of consonant-vowel (CV) units under background noise using combined temporal and spectral preprocessing. Int. J. Speech Tech. (Springer) **14**(3), 259–272 (2011)

113. A.K. Vuppala, K.S. Rao, S. Chakrabarti, Spotting and recognition of consonant-vowel units from continuous speech using accurate vowel onset points. Circ. Syst. Signal Process. (Springer) **31**(4), 1459–1474 (2012)

114. A.K. Vuppala, K.S. Rao, S. Chakrabarti, Improved speaker identification in wireless environment. Int. J. Signal Imag. Syst. Eng. **6**(3), 130–137 (2013)

115. A.K. Vuppala, K.S. Rao, Speaker identification under background noise using features extracted from steady vowel regions. Wiley Int. J. Adapt. Contr. Signal Process. **29**, 781–792 (2013)

116. A.K. Vuppala, S. Chakrabarti, K.S. Rao, L. Dutta, "Robust speaker recognition on mobile devices," in *Proc. of IEEE Int. Conf. on Signal Processing and Communications* (Bangalore, India, 2010)

117. K.S. Prahallad, B. Yegnanarayana, S.V. Gangashetty, Online text-independent speaker verification system using autoassociative neural network models, in *Proc. of INNS-IEEE Int. Joint Conf. Neural Networks* (Washington DC, USA, 2001), pp. 1548–1553

118. B. Yegnanarayana, S.P. Kishore, AANN an alternative to GMM for pattern recognition. Neural Network **15**, 459–469 (2002)

119. A.K. Vuppala, S. Chakrabarti, K.S. Rao, Effect of speech coding on speaker identification, in *Proc. of IEEE INDICON* (Kolkata, India, 2010)

120. S. Sigurdsson, K.B. Petersen, T. Lehn-Schioler, Mel frequency cepstral coefficients: An evaluation of robustness of MP3 encoded music, in *Proc. of Seventh Int. Conf. on Music Information Retrieval*, 2006

121. A.L. Edwards, *An Introduction to Linear Regression and Correlation* (W.H. Freeman and Company Ltd, Cranbury, NJ, USA, 1976)

122. J.R. Deller, J.G. Proakis, J.H.L. Hansen, *Discrete-Time Processing of Speech Signals* (Macmilan Publishing, New York, 1993)

123. R.V. Hogg, J. Ledolter, *Engineering Statistics* (Macmillan Publishing, New York, 1987)

124. S.V. Gangashetty, C.C. Sekhar, B. Yegnanarayana, Detection of vowel onset points in continuous speech using autoassociative neural network models, in *Proc. Int. Conf. Spoken Language Processing*, pp. 401–410, 2004

125. J.R. Deller, J.H. Hansen, J.G. Proakis, *Discrete Time Processing of Speech Signals*, 1st edn. (Prentice Hall PTR, Upper Saddle River, NJ, 1993)

126. J. Benesty, M.M. Sondhi, Y.A. Huang, *Springer Handbook of Speech Processing* (Springer, New York, 2008)

127. J. Volkmann, S. Stevens, E. Newman, A scale for the measurement of the psychological magnitude pitch. J. Acoust. Soc. Am. **8**, 185–190 (1937)

128. Z. Fang, Z. Guoliang, S. Zhanjiang, Comparison of different implementations of MFCC. J. Comput. Sci. Tech. **16**(6), 582–589 (2001)

129. G.K.T. Ganchev, N. Fakotakis, Comparative evaluation of various MFCC implementations on the speaker verification task, in *Proc. of Int. Conf. on Speech and Computer* (Patras, Greece, 2005), pp. 191–194

130. L.R. Rabiner, B.H. Juang, *Fundamentals of speech Recognition* (Prentice Hall PTR, Englewood cliffs, NJ, 1993)

131. S. Furui, Comparison of speaker recognition methods using statistical features and dynamic features. IEEE Trans. Acoust. Speech Signal Process. **29**(3), 342–350 (1981)

132. J.S. Mason, X. Zhang, Velocity and acceleration features in speaker recognition, in *Proc. IEEE Int. Conf. Acoust., Speech, Signal Processing*, (Toronto, Canada, 1991), pp. 3673–3676

133. W.C. Chu, *Speech Coding Algorithms: Foundation and Evolution of Standardized Coders* (Wiley, New York, 2003)

134. A.M. Kondoz, *Digital Speech: Coding for Low Bit Rate Communication Systems, 2nd edn.* (Wiley, New York, 2004)

135. H.L.J. Hansen, B.L. Pellom, An effective quality evaluation protocol for speech enhancement algorithm, in *Proc. Int. Conf. Spoken Language Processing*, pp. 2819–2822, 1998

136. L.R. Rabiner, A tutorial on hidden Markov models and selected applications in speech recognition, in *Proc. of IEEE*, pp. 257–286, 1989

137. S. Theodoridis, K. Koutroumbas, *Pattern Recognition*, 3rd edn. (Elsevier, Academic Press, Waltham, MA, USA, 2006)